Muhittin E. Aydin, Svetlin G. Georgiev

Differential Geometry

Also of Interest

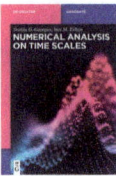

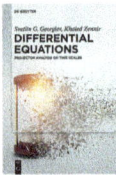

Muhittin E. Aydin, Svetlin G. Georgiev

Differential Geometry

—

Frenet Equations and Differentiable Maps

DE GRUYTER

Mathematics Subject Classification 2020
Primary: 53-01, 53A04, 53A05; Secondary: 53A10

Authors

Dr. Muhittin E. Aydin
Firat University
Department of Mathematics
23200 Elazig
Turkey
meaydin@firat.edu.tr

Prof. Dr. Svetlin G. Georgiev
Kliment Ohridski University of Sofia
Department of Differential Equations
Faculty of Mathematics and Informatics
1126 Sofia
Bulgaria
svetlingeorgiev1@gmail.com

ISBN 978-3-11-150089-8
e-ISBN (PDF) 978-3-11-150185-7
e-ISBN (EPUB) 978-3-11-150223-6

Library of Congress Control Number: 2024938449

Bibliographic information published by the Deutsche Nationalbibliothek
The Deutsche Nationalbibliothek lists this publication in the Deutsche Nationalbibliografie;
detailed bibliographic data are available on the Internet at http://dnb.dnb.de.

© 2024 Walter de Gruyter GmbH, Berlin/Boston
Cover image: piranka / iStock / Getty Images Plus
Typesetting: VTeX UAB, Lithuania
Printing and binding: CPI books GmbH, Leck

www.degruyter.com

Preface

Differential geometry is the study of curves, surfaces, and higher-dimensional objects by using techniques from linear algebra and calculus. But still, average knowledge of linear algebra and (multivariable) calculus is sufficient so that readers can follow the results of this book.

Once we decided to write this book, we wanted to organize it in a different format from the many books that already exist in the literature on differential geometry. In view of the fact that a mathematical subject can only be learned through much practice rather than reading directly, we based the book on examples, exercises, and problems. In this sense, we give around 300 advanced practical problems, which are placed at the end of each chapter. Although complete solutions to these problems are provided in the book, we encourage the reader to solve them with his or her own efforts. Just in case they require to check the solutions, readers should consult our solutions. Furthermore, the book contains over 45 illustrations that provide the opportunity to better visualize the theoretical concepts.

The book, however, addresses many notable classical results such as Lancret, Shell, Joachimsthal, and Meusnier theorems, as well as the fundamental theorems of plane curves, space curves, surfaces, and manifolds.

We hope that the format of the present book will contribute to the reader's knowledge of differential geometry.

https://doi.org/10.1515/9783111501857-201

Contents

1 Curves in $\mathbb{R}^n$

In this chapter we first consider the concept of Frenet curve and then define its tangent line, normal (hyper)plane and osculating plane. After introducing the famous Frenet formulas, curvature, curvature vector, and torsion are introduced as applications. In terms of these basic invariants, the local behavior of a parameterized curve around biregular points is observed. The well-known Lancret theorem for general helices and Shell theorem for Bertrand curves are revisited. A rigid motion of a Euclidean space is given so that the existence and uniqueness theorems can be stated.

1.1 Frenet curves in $\mathbb{R}^n$

Let $a, b \in \mathbb{R}$, $a < b$.

Definition 1.1. A curve in $\mathbb{R}^n$ is a continuous function $f : [a, b] \to \mathbb{R}^n$. The function f that defines the curve is called a parameterization of the curve and the curve is a parametric curve.

Example 1.1. Suppose that $f : [1, 2] \to \mathbb{R}^3$ is given by

$$f(t) = (t - 1, t, t^2 - 5), \quad t \in [1, 2].$$

Then $f : [1, 2] \to \mathbb{R}^3$ is a curve. Its graph is shown in Fig. 1.1.

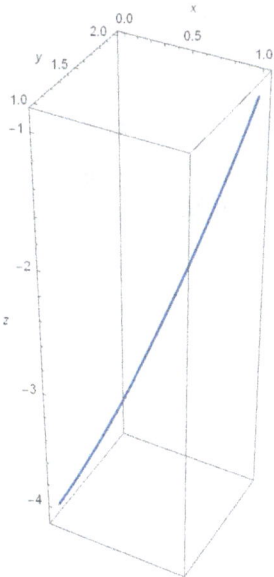

Figure 1.1: A parametric curve in $\mathbb{R}^3$ given by $f(t) = (t - 1, t, t^2 - 5)$, $t \in [1, 2]$.

https://doi.org/10.1515/9783111501857-001

Definition 1.2. A regular parameterized curve is a function $f : [a,b] \to \mathbb{R}^n$ such that $f \in C^1([a,b])$ and

$$f'(t) \ne (0,0,\ldots,0) \quad \text{for every } t \in [a,b].$$

Definition 1.3. Let $f : [a,b] \to \mathbb{R}^n$ be a parameterized curve such that $f \in C^1([a,b])$. Then the vector $f'(t_0)$ is called the tangent vector to f at t_0 and the line spanned by this vector and passing through $f(t_0)$ is called the tangent line to f at this point.

Example 1.2. Let $[a,b] = [0,1]$ and $f : [0,1] \to \mathbb{R}^2$ be given by

$$f(t) = (t, t^3 + t^2), \quad t \in [0,1].$$

Here

$$f_1(t) = t,$$
$$f_2(t) = t^3 + t^2, \quad t \in [0,1].$$

Hence,

$$f_1'(t) = 1,$$
$$f_2'(t) = 3t^2 + 2t, \quad t \in [0,1].$$

Thus,

$$f'(t) = (f_1'(t), f_2'(t))$$
$$= (1, 3t^2 + 2t)$$
$$\ne (0,0) \quad \text{for every } t \in [0,1].$$

Thus, the considered curve is a regular curve; see Fig. 1.2. In addition, the tangent line parallel to the tangent vector $f'(t_0)$ and passing through $f(t_0)$, $t_0 \in [0,1]$, is

$$(t_0, t_0^3 + t_0^2) + \lambda(1, 3t_0^2 + 2t_0),$$

where λ is a parameter.

Example 1.3. Let $f : [0,\pi] \to \mathbb{R}^2$ be given by

$$f(t) = \frac{1}{2}(\cos(2t), \sin(2t)), \quad t \in [0,\pi].$$

This curve is a circle of radius $\frac{1}{2}$ centered at $(0,0)$. Here

$$f_1(t) = \frac{1}{2}\cos(2t),$$

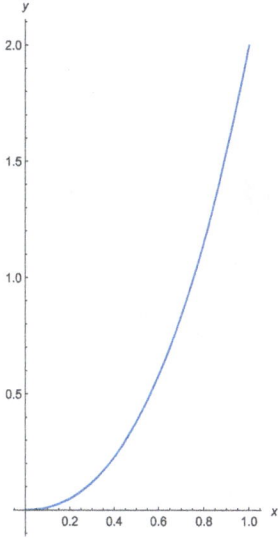

Figure 1.2: A regular parameterized curve in $\mathbb{R}^2$ given by $f(t) = (t, t^3 + t^2)$, $t \in [0, 1]$.

$$f_2(t) = \frac{1}{2}\sin(2t), \quad t \in [0, \pi].$$

Then

$$f_1'(t) = -\sin(2t),$$
$$f_2'(t) = \cos(2t), \quad t \in [0, \pi].$$

Hence,

$$f'(t) = (f_1'(t), f_2'(t))$$
$$= (-\sin(2t), \cos(2t))$$
$$\neq (0, 0) \quad \text{for every } t \in [0, \pi].$$

Thus, the considered curve is a regular curve (see Fig. 1.3). In addition, the tangent line parallel to the tangent vector $f'(t_0)$ and passing through $f(t_0)$, $t_0 \in [0, \pi]$, is

$$\frac{1}{2}(\cos(2t_0), \sin(2t_0)) + \lambda(-\sin(2t_0), \cos(2t_0)),$$

where λ is a parameter.

Example 1.4. Let $f : [0, 4\pi] \to \mathbb{R}^3$ be given by

$$f(t) = (\cos t, \sin t, t), \quad t \in [0, 4\pi].$$

This curve is called a (circular) helix. Here

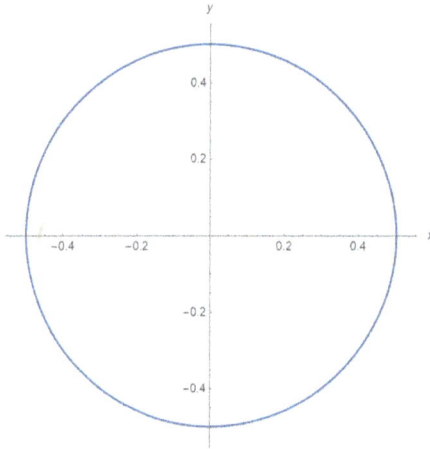

Figure 1.3: A circle in $\mathbb{R}^2$ of radius $1/2$ centered at $(0,0)$. A regular parameterization is $f(t) = (\cos(2t)/2, \sin(2t)/2), t \in [0, \pi]$.

$$f_1(t) = \cos t,$$
$$f_2(t) = \sin t,$$
$$f_3(t) = t, \quad t \in [0, 4\pi].$$

Then

$$f_1'(t) = -\sin t,$$
$$f_2'(t) = \cos t,$$
$$f_3'(t) = 1, \quad t \in [0, 4\pi].$$

Hence,

$$f'(t) = (f_1'(t), f_2'(t), f_3'(t))$$
$$= (-\sin t, \cos t, 1)$$
$$\neq (0, 0, 0) \quad \text{for every } t \in [0, 4\pi].$$

Thus, the considered curve is a regular curve and drawn in Fig. 1.4. In addition, the tangent line parallel to the tangent vector $f'(t_0)$ and passing through $f(t_0)$, $t_0 \in [0, 4\pi]$, is

$$(\cos t_0, \sin t_0, t_0) + \lambda(-\sin t_0, \cos t_0, 1),$$

where λ is a parameter.

Exercise 1.1. Prove that the following curves are regular:
1. For $n = 2$, $[a, b] = [-10, 10]$,

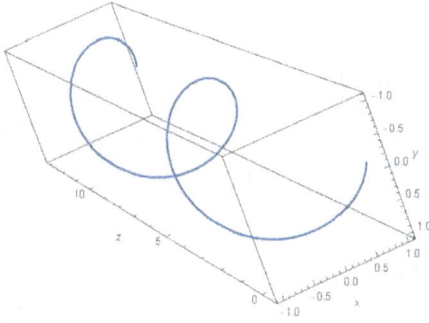

Figure 1.4: A circular helix in $\mathbb{R}^3$ parameterized by $f(t) = (\cos t, \sin t, t)$, $t \in [0, 4\pi]$.

$$f(t) = (2t, -4t), \quad t \in [-10, 20].$$

2. For $n = 3$, $[a, b] = [0, 2\pi]$,

$$f(t) = (2\cos(3t), 3\sin(4t), 5t), \quad t \in [0, 2\pi].$$

3. For $n = 2$, $t \in [1, 10]$,

$$f(t) = (t^2, t^3), \quad t \in [1, 10].$$

Exercise 1.2. Let $u_2(t), \ldots, u_n(t) \in C^1([a, b])$. Prove that

$$f(t) = (t, u_2(t), \ldots, u_n(t)), \quad t \in [a, b],$$

is always a regular parameterized curve in $\mathbb{R}^n$.

Definition 1.4. Let $[\alpha, \beta] \subset \mathbb{R}$. Suppose $\phi : [\alpha, \beta] \to [a, b]$ is such that $\phi \in C^1([\alpha, \beta])$ and $\phi'(t) > 0$, for every $t \in [\alpha, \beta]$. Let also $f : [a, b] \to \mathbb{R}^n$ be a regular parameterized curve. Then the curves f and $f \circ \phi$ are said to be equivalent.

Example 1.5. Let $f : [0, \frac{2}{3}] \to \mathbb{R}^2$ be given by

$$f(t) = (1 + t, t^2), \quad t \in \left[0, \frac{2}{3}\right].$$

Here

$$f_1(t) = 1 + t,$$

$$f_2(t) = t^2, \quad t \in \left[0, \frac{2}{3}\right].$$

Then

$$f_1'(t) = 1,$$

$$f_2'(t) = 2t, \quad t \in \left[0, \frac{2}{3}\right],$$

and

$$f'(t) = (f_1'(t), f_2'(t))$$

$$= (1, 2t), \quad t \in \left[0, \frac{2}{3}\right],$$

i. e., $f : [0, \frac{2}{3}] \to \mathbb{R}^2$ is a regular curve. Let also $\phi : [0, 2] \to [0, \frac{2}{3}]$ be given by

$$\phi(s) = \frac{s}{s+1}, \quad s \in [0, 2].$$

Then $\phi : [0, 2] \to [0, \frac{2}{3}]$ and

$$\phi'(s) = \frac{s + 1 - s}{(s + 1)^2}$$

$$= \frac{1}{(s + 1)^2} > 0 \quad \text{for every } s \in [0, 2].$$

Furthermore,

$$f \circ \phi(s) = \left(1 + \frac{s}{1 + s}, \frac{s^2}{(1 + s)^2}\right)$$

$$= \left(\frac{2s + 1}{s + 1}, \frac{s^2}{(s + 1)^2}\right), \quad s \in [0, 2].$$

Therefore, f and $f \circ \phi$ are equivalent; see Fig. 1.5.

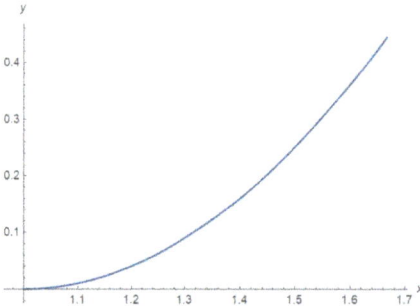

Figure 1.5: A parabola in $\mathbb{R}^2$ of the form $y = (x - 1)^2$. The equivalent parameterizations are $f(t) = (t + 1, t^2)$, $t \in [0, 2/3]$, and $f \circ \phi(s) = ((2s + 1)/(s + 1), s^2/(s + 1)^2)$, $s \in [0, 2]$.

Example 1.6. Consider the Diocles cissoid

$$f(t) = \left(\frac{a}{t^2 + 1}, \frac{a}{t(t^2 + 1)} \right), \quad t \in \left[\frac{1}{2}, 1 \right],$$

where $a > 0$ is a given parameter. Here

$$f_1(t) = \frac{a}{t^2 + 1},$$

$$f_2(t) = \frac{a}{t(t^2 + 1)}, \quad t \in \left[\frac{1}{2}, 1 \right].$$

Then

$$f_1'(t) = -\frac{2at}{(t^2 + 1)^2},$$

$$f_2'(t) = -\frac{a(3t^2 + 1)}{t^2(t^2 + 1)^2}, \quad t \in \left[\frac{1}{2}, 1 \right].$$

Hence,

$$f'(t) = (f_1'(t), f_2'(t))$$

$$\neq (0,0) \quad \text{for every } t \in \left[\frac{1}{2}, 1 \right],$$

i. e., $f : [\frac{1}{2}, 1] \to \mathbb{R}^2$ is a regular curve. Let

$$\phi(s) = -\frac{1}{s}, \quad s \in [-2, -1].$$

Then $\phi : [-2, -1] \to [\frac{1}{2}, 1]$ and

$$\phi'(s) = \frac{1}{s^2} > 0 \quad \text{for every } s \in [-2, -1].$$

Also

$$f \circ \phi(s) = f(\phi(s))$$

$$= \left(\frac{a}{(-1/s)^2 + 1}, \frac{a}{(-1/s)((-1/s)^2 + 1)} \right)$$

$$= \left(\frac{as^2}{s^2 + 1}, \frac{-as^3}{s^2 + 1} \right), \quad s \in [-2, -1].$$

We have that f and $f \circ \phi$ are equivalent curves. For the value $a = 1$, see Fig. 1.6.

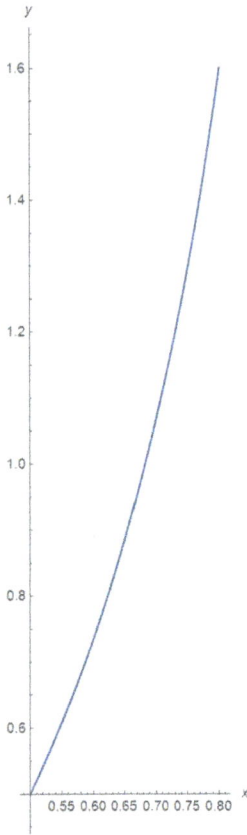

Figure 1.6: Diocles cissoid in $\mathbb{R}^2$ of the form $x^3 + xy^2 - y^2 = 0$. The equivalent parameterizations are $f(t) = (1/(t^2 + 1), 1/(t^3 + t))$, $t \in [1/2, 1]$, and $f \circ \phi(s) = (s^2/(s^2 + 1), -s^3/(s^2 + 1))$, $s \in [-2, -1]$.

Example 1.7. Consider the witch of Maria Agnesi

$$f(t) = (a \cot t, a(\sin t)^2), \quad t \in \left[\frac{\pi}{4}, \frac{\pi}{2}\right],$$

where $a > 0$ is given parameter. Here

$$f_1(t) = a \cot t,$$

$$f_2(t) = a(\sin t)^2, \quad t \in \left[\frac{\pi}{4}, \frac{\pi}{2}\right].$$

Then

$$f_1'(t) = -a(1 + \cot^2 t),$$

$$f_2'(t) = a \sin(2t), \quad t \in \left[\frac{\pi}{4}, \frac{\pi}{2}\right],$$

and

$$f'(t) = (f_1'(t), f_2'(t))$$
$$= (-a(1 + \cot^2 t), a\sin(2t))$$
$$\neq (0, 0), \quad t \in \left[\frac{\pi}{4}, \frac{\pi}{2}\right],$$

i.e., $f : [\frac{\pi}{4}, \frac{\pi}{2}] \to \mathbb{R}^2$ is a regular curve. Let

$$\phi(s) = s - \frac{\pi}{2}, \quad s \in \left[\frac{3\pi}{4}, \pi\right].$$

Then $\phi : [\frac{3\pi}{4}, \pi] \to [\frac{\pi}{4}, \frac{\pi}{2}]$ and

$$\phi'(s) = 1 > 0 \quad \text{for every } s \in \left[\frac{3\pi}{4}, \pi\right],$$

and

$$f \circ \phi(s) = f(\phi(s))$$
$$= \left(a\cot\left(s - \frac{\pi}{2}\right), a\left(\sin\left(s - \frac{\pi}{2}\right)\right)^2 \right)$$
$$= (a\tan s, a(\cos s)^2), \quad s \in \left[\frac{3\pi}{4}, \pi\right].$$

We have that f and $f \circ \phi$ are equivalent curves. For the value $a = 1$, see Fig. 1.7.

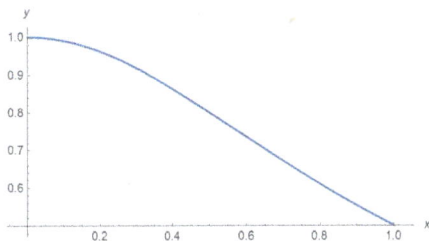

Figure 1.7: Witch of Maria Agnesi in $\mathbb{R}^2$. The equivalent parameterizations are $f(t) = (\cot t, (\sin t)^2), t \in [\pi/4, \pi/2]$, and $f \circ \phi(s) = (\tan s, (\cos s)^2), s \in [3\pi/4, \pi]$.

Exercise 1.3. Prove that the following curves are equivalent:
1. (strophoid)

$$f(t) = \left(\frac{2at^2}{1 + t^2}, \frac{a + (t^2 - 1)}{1 + t^2} \right), \quad t \in \left[\frac{1}{16}, \frac{1}{4}\right],$$

and

$$g(s) = \left(\frac{2as^4}{1+s^4}, \frac{a+(s^4-1)}{1+s^4} \right), \quad s \in \left[\frac{1}{4}, \frac{1}{2} \right];$$

2. (astroid)

$$f(t) = (a(\cos t)^3, a(\sin t)^3), \quad t \in \left[\frac{\pi}{4}, \frac{\pi}{2} \right],$$

and

$$g(s) = \left(a(1-s^2)^{\frac{3}{2}}, as^3 \right), \quad s \in \left[\frac{\sqrt{2}}{2}, 1 \right];$$

3. (cycloid)

$$f(t) = (a(t - \sin t), a(1 - \cos t)), \quad t \in \left[\frac{\pi}{4}, \frac{\pi}{2} \right],$$

and

$$g(s) = \left(a(\arcsin s - s), a(1 - \sqrt{1-s^2}) \right), \quad s \in \left[\frac{\sqrt{2}}{2}, 1 \right].$$

Hint 1.1. Use the following functions:
1. $\phi(s) = s^2, s \in [\frac{1}{4}, \frac{1}{2}]$;
2. $\phi(s) = \arcsin s, s \in [\frac{\sqrt{2}}{2}, 1]$;
3. $\phi(s) = \arcsin s, s \in [\frac{\sqrt{2}}{2}, 1]$.

Definition 1.5. Let $f : [a, b] \to \mathbb{R}^n$ be a regular parameterized curve and

$$f(t) = (f_1(t), f_2(t), \ldots, f_n(t)), \quad t \in [a, b].$$

The arc length parameter $L_f(t, a)$, $t \in [a, b]$, is defined as follows:

$$L_f(t, a) = \int_a^t |f'(s)| ds.$$

Example 1.8. Consider the circle given in Example 1.3. Using the computations there, we get

$$\begin{aligned} |f'(t)| &= \sqrt{(f_1'(t))^2 + (f_2'(t))^2} \\ &= \sqrt{(-\sin(2t))^2 + (\cos(2t))^2} \\ &= 1, \quad t \in [0, \pi]. \end{aligned}$$

Hence,

$$L_f(t,0) = \int_0^t |f'(s)|ds$$

$$= \int_0^t ds$$

$$= t, \quad t \in [0, \pi].$$

Example 1.9. Consider the circular helix given in Example 1.4. Using the computations there, we find

$$|f'(t)| = \sqrt{(f_1'(t))^2 + (f_2'(t))^2 + (f_3'(t))^2}$$

$$= \sqrt{(-\sin t)^2 + (\cos t)^2 + 1}$$

$$= \sqrt{2}, \quad t \in [0, 4\pi].$$

Then

$$L_f(t,0) = \int_0^t |f'(s)|ds$$

$$= \int_0^t \sqrt{2}ds$$

$$= \sqrt{2}t, \quad t \in [0, 4\pi].$$

Example 1.10. Consider the parabola given in Example 1.5. Using the computations there, we find

$$|f'(t)| = \sqrt{(f_1'(t))^2 + (f_2'(t))^2}$$

$$= \sqrt{1 + 4t^2}, \quad t \in \left[0, \frac{2}{3}\right].$$

Hence,

$$L_f(t,0) = \int_0^t |f'(s)|ds$$

$$= \int_0^t \sqrt{1 + 4s^2}ds, \quad t \in \left[0, \frac{2}{3}\right].$$

Note that

$$\int \sqrt{1+4y^2}\,dy = \frac{1}{2}y\sqrt{1+4y^2} + \frac{1}{4}\log(2y + \sqrt{1+4y^2}) + c, \quad y \in \mathbb{R},$$

where c is a constant. Hence,

$$L_f(t,0) = \left(\frac{1}{2}y\sqrt{1+4y^2}\Big|_{y=0}^{y=t} + \frac{1}{4}\log(2y + \sqrt{1+4y^2})\Big|_{y=0}^{y=t} \right)$$

$$= \left(\frac{1}{2}t\sqrt{1+4t^2} + \frac{1}{4}\log(2t + \sqrt{1+4t^2}) \right), \quad t \in \left[0, \frac{2}{3} \right].$$

Exercise 1.4. Find the arc length of the following curves:

1.
$$f(t) = (t, t^{\frac{3}{2}}), \quad t \in [a,b];$$

2.
$$f(t) = (t, t^2), \quad t \in [a,b];$$

3.
$$f(t) = (t, \log t), \quad t \in [a,b];$$

4.
$$f(t) = \left(t, c\cosh\left(\frac{t}{c}\right) \right), \quad t \in [a,b],$$

where $c > 0$ is a given constant;

5.
$$f(t) = (t, e^t), \quad t \in [a,b].$$

Answer 1.2. 1.
$$L_f(t,a) = \frac{1}{27}\left((4+9t)^{\frac{3}{2}} - (4+9a)^{\frac{3}{2}} \right), \quad t \in [a,b].$$

2.
$$L_f(t,a) = \frac{1}{2}(t\sqrt{1+t^2} - a\sqrt{1+a^2}) + \frac{1}{4}\log\frac{2t + \sqrt{1+4t^2}}{2a + \sqrt{1+4a^2}}, \quad t \in [a,b].$$

3.
$$L_f(t,a) = \sqrt{1+t^2} - \sqrt{1+a^2}$$
$$- \frac{1}{2}\left(\log\frac{1+\sqrt{1+t^2}}{1-\sqrt{1+t^2}} - \log\frac{1+\sqrt{1+a^2}}{1-\sqrt{1+a^2}} \right), \quad t \in [a,b].$$

4.
$$L_f(t,a) = c\left(\sinh\left(\frac{t}{c}\right) - \sinh\left(\frac{a}{c}\right) \right), \quad t \in [a,b].$$

5.
$$L_f(t,a) = \sqrt{1+e^{2t}} - \sqrt{1+e^{2a}}$$
$$- \frac{1}{2}\left(\log\frac{1+e^{2t}}{1-\sqrt{1+e^{2t}}} - \log\frac{1+\sqrt{1+e^{2a}}}{1-\sqrt{1+e^{2a}}} \right), \quad t \in [a,b].$$

Definition 1.6. We will say that a regular parameterized curve is naturally parameterized if

$$|f'(s)| = 1, \quad s \in [a, b].$$

Usually, the natural parameter is denoted by s.

Exercise 1.5. Prove that the arc lengths of any two equivalent curves are equal.

Solution. Let $f : [a, b] \to \mathbb{R}^n$ be a regular curve, $\phi : [\alpha, \beta] \to [a, b]$, $\phi(\alpha) = a$, $\phi(\beta) = b$, $\phi \in C^1([\alpha, \beta])$, $\phi' > 0$ on $[\alpha, \beta]$. Then f and $f \circ \phi$ are equivalent. Set

$$g = f \circ \phi.$$

Then

$$L_g(\beta, \alpha) = \int_\alpha^\beta |g'(\tau)| d\tau$$

$$= \int_\alpha^\beta |f'(\phi(\tau))\phi'(\tau)| d\tau$$

$$= \int_\alpha^\beta |f'(\phi(\tau))||\phi'(\tau)| d\tau$$

$$= \int_\alpha^\beta |f'(\phi(\tau))| d\phi(\tau)$$

$$= \int_a^b |f'(t)| dt$$

$$= L_f(b, a).$$

This completes the proof.

Exercise 1.6. Prove that for any regular parameterized curve there is a naturally parameterized curve that is equivalent to it.

Solution. Let $f : [a, b] \to \mathbb{R}^n$ be a regular parameterized curve. Define

$$\phi(t) = \int_a^t |f'(\tau)| d\tau, \quad t \in [a, b].$$

We have that $\phi \in C^1([a, b])$ and

$$\phi'(t) = |f'(t)| > 0, \quad t \in [a, b].$$

Hence, $\phi' > 0$ on $[a, b]$ and $\phi : [a, b] \to [\alpha, \beta]$ is a diffeomorphism for some interval $[\alpha, \beta] \subset \mathbb{R}$. Note that

$$\begin{aligned}
(\phi^{-1})'(s) &= \frac{1}{\phi'(\phi^{-1}(s))} \\
&= \frac{1}{|f'(\phi^{-1}(s))|}, \quad s \in [\alpha, \beta].
\end{aligned}$$

Let

$$g(s) = f(\phi^{-1}(s)), \quad s \in [\alpha, \beta].$$

Then

$$\begin{aligned}
g'(s) &= f'(\phi^{-1}(s))(\phi^{-1})'(s) \\
&= \frac{f'(\phi^{-1}(s))}{|f'(\phi^{-1}(s))|}, \quad s \in [\alpha, \beta],
\end{aligned}$$

and

$$|g'(s)| = 1, \quad s \in [\alpha, \beta].$$

This completes the proof.

Remark 1.1. Note that we have

$$f(t) = g(s(t)), \quad t \in [a, b],$$

and

$$\begin{aligned}
f'(t) &= g'(s(t))s'(t) \\
&= g'(s(t))|f'(t)|, \quad t \in [a, b].
\end{aligned}$$

Example 1.11. Consider the curve

$$f(t) = (at^3, bt^3), \quad t \in [2, 10],$$

where $a, b \in \mathbb{R}$ are such that $a^2 + b^2 \neq 0$. Here

$$\begin{aligned}
f_1(t) &= at^3, \\
f_2(t) &= bt^3, \quad t \in [2, 10].
\end{aligned}$$

Then

$$f_1'(t) = 3at^2,$$
$$f_2'(t) = 3bt^2, \quad t \in [2, 10],$$

and

$$|f'(t)| = \sqrt{(f_1'(t))^2 + (f_2'(t))^2}$$
$$= \sqrt{(3at^2) + (3bt^2)^2}$$
$$= \sqrt{9a^2t^4 + 9b^2t^4}$$
$$= 3t^2 \sqrt{a^2 + b^2}, \quad t \in [2, 10].$$

Thus, $f : [2, 10] \rightarrow \mathbb{R}^2$ is a regular curve that is not naturally parameterized. For its arc length, we have

$$L_f(t, 2) = -\int_2^t |f'(s)| ds$$

$$= 3\sqrt{a^2 + b^2} \int_2^t s^2 ds$$

$$= 3\sqrt{a^2 + b^2} \left. \frac{s^3}{3} \right|_{s=2}^{s=t}$$

$$= \sqrt{a^2 + b^2}(t^3 - 8), \quad t \in [2, 10].$$

Now set $L_f(t, 2) = \phi(t)$ and $s = \phi(t)$ where $s \in [0, 992\sqrt{a^2 + b^2}]$. Hence

$$\frac{1}{\sqrt{a^2 + b^2}} s = t^3 - 8, \qquad s \in [0, 992\sqrt{a^2 + b^2}],$$

$$t^3 = \frac{1}{\sqrt{a^2 + b^2}} s + 8, \quad s \in [0, 992\sqrt{a^2 + b^2}],$$

and

$$t = \sqrt[3]{\frac{1}{\sqrt{a^2 + b^2}} s + 8}, \quad s \in [0, 992\sqrt{a^2 + b^2}].$$

Therefore,

$$\phi^{-1}(s) = \sqrt[3]{\frac{1}{\sqrt{a^2 + b^2}} s + 8}, \quad s \in [0, 992\sqrt{a^2 + b^2}].$$

Consider the curve

$$g(s) = f \circ \phi^{-1}(s), \quad s \in [0, 992\sqrt{a^2 + b^2}].$$

Then

$$g(s) = \left(a(\phi^{-1}(s))^3, b(\phi^{-1}(s))^3\right)$$
$$= \left(a\left(\frac{1}{\sqrt{a^2 + b^2}}s + 8\right), b\left(\frac{1}{\sqrt{a^2 + b^2}}s + 8\right)\right), \quad s \in [0, 992\sqrt{a^2 + b^2}].$$

Note that

$$g_1(s) = a\left(\frac{1}{\sqrt{a^2 + b^2}}s + 8\right),$$

$$g_2(s) = b\left(\frac{1}{\sqrt{a^2 + b^2}}s + 8\right), \quad t \in [0, 992\sqrt{a^2 + b^2}],$$

and

$$g_1'(s) = \frac{a}{\sqrt{a^2 + b^2}},$$

$$g_2'(s) = \frac{b}{\sqrt{a^2 + b^2}}, \quad t \in [0, 992\sqrt{a^2 + b^2}].$$

Then

$$|g'(s)| = \sqrt{\left(g_1'(s)\right)^2 + \left(g_2'(s)\right)^2}$$
$$= \sqrt{\frac{a^2}{a^2 + b^2} + \frac{b^2}{a^2 + b^2}}$$
$$= 1, \quad s \in [0, 992\sqrt{a^2 + b^2}].$$

Thus, $g : [0, 992\sqrt{a^2 + b^2}] \to \mathbb{R}^2$ is a naturally parameterized curve that is equivalent to the curve f.

Example 1.12. Consider the circle

$$f(t) = \left(\frac{1}{a}\cos(at), \frac{1}{a}\sin(at)\right), \quad t \in [0, 2\pi].$$

Here

$$f_1(t) = \frac{1}{a}\cos(at),$$

$$f_2(t) = \frac{1}{a}\sin(at), \quad t \in [0, 2\pi].$$

Then

$$f_1'(t) = -\sin(at),$$
$$f_2'(t) = \cos(at), \quad t \in [0, 2\pi],$$

and

$$|f'(t)| = \sqrt{(f_1'(t))^2 + (f_2'(t))^2}$$
$$= \sqrt{(\sin(at))^2 + (\cos(at))^2}$$
$$= 1, \quad t \in [0, 2\pi].$$

Thus, $f : [0, 2\pi] \to \mathbb{R}^2$ is a naturally parameterized curve.

Example 1.13. Consider the circular helix

$$f(t) = (c\cos(at), c\sin(at), \beta t), \quad t \in [a, b],$$

where $a, \beta, c \in \mathbb{R}$ are such that $a^2 c^2 + \beta^2 \neq 0$. Here

$$f_1(t) = c\cos(at),$$
$$f_2(t) = c\sin(at),$$
$$f_3(t) = \beta t, \quad t \in [a, b].$$

Then

$$f_1'(t) = -ca\sin(at),$$
$$f_2'(t) = ca\sin(at),$$
$$f_3'(t) = \beta, \quad t \in [a, b],$$

and

$$|f'(t)| = \sqrt{(f_1'(t))^2 + (f_2'(t))^2 + (f_3'(t))^2}$$
$$= \sqrt{(-ca\sin(at))^2 + (ca\cos(at))^2 + \beta^2}$$
$$= \sqrt{c^2 a^2(\sin(at))^2 + c^2 a^2(\cos(at))^2 + \beta^2}$$
$$= \sqrt{c^2 a^2 + \beta^2}, \quad t \in [a, b].$$

Thus, $f : [a, b] \to \mathbb{R}^3$ is a regular curve that is not naturally parameterized. For its arc length, we have

$$L_f(t, a) = \int_a^t |f'(s)| ds$$

$$= \int_a^t \sqrt{c^2 a^2 + \beta^2} \, ds$$

$$= \sqrt{c^2 a^2 + \beta^2}(t - a), \quad t \in [a, b],$$

whereupon

$$t - a = \frac{1}{\sqrt{c^2 a^2 + \beta^2}} L_f(t, a), \quad t \in [a, b].$$

Now set $L_f(t, 2) = \phi(t)$ and $s = \phi(t)$ where $s \in [0, (b - a)\sqrt{c^2 a^2 + \beta^2}]$. Hence,

$$\phi^{-1}(s) = a + \frac{1}{\sqrt{c^2 a^2 + \beta^2}} s, \quad s \in [0, (b - a)\sqrt{c^2 a^2 + \beta^2}].$$

Consider the curve

$$g(s) = f \circ \phi^{-1}(s), \quad s \in [0, (b - a)\sqrt{c^2 a^2 + \beta^2}].$$

We have

$$g(s) = \left(c \cos\left(a\left(a + \frac{1}{\sqrt{c^2 a^2 + \beta^2}} s \right) \right), c \sin\left(a\left(a + \frac{1}{\sqrt{c^2 a^2 + \beta^2}} s \right) \right),$$

$$\beta\left(a + \frac{1}{\sqrt{c^2 a^2 + \beta^2}} s \right) \right), \quad s \in [0, (b - a)\sqrt{c^2 a^2 + \beta^2}].$$

Here

$$g_1(s) = c \cos\left(a\left(a + \frac{1}{\sqrt{c^2 a^2 + \beta^2}} s \right) \right),$$

$$g_2(s) = c \sin\left(a\left(a + \frac{1}{\sqrt{c^2 a^2 + \beta^2}} s \right) \right),$$

$$g_3(s) = \beta\left(a + \frac{1}{\sqrt{c^2 a^2 + \beta^2}} s \right), \quad s \in [0, (b - a)\sqrt{c^2 a^2 + \beta^2}].$$

Then

$$g_1'(s) = -\frac{ca}{\sqrt{c^2 a^2 + \beta^2}} \sin\left(a\left(a + \frac{1}{\sqrt{c^2 a^2 + \beta^2}} s \right) \right),$$

$$g_2'(s) = \frac{ca}{\sqrt{c^2 a^2 + \beta^2}} \cos\left(a\left(a + \frac{1}{\sqrt{c^2 a^2 + \beta^2}} s \right) \right),$$

$$g_3(s) = \frac{\beta}{\sqrt{c^2\alpha^2 + \beta^2}}, \quad s \in [0, (b-a)\sqrt{c^2\alpha^2 + \beta^2}],$$

and

$$|g'(s)| = \left(\left(-\frac{c\alpha}{\sqrt{c^2\alpha^2 + \beta^2}}\cos\left(\alpha\left(a + \frac{1}{\sqrt{c^2\alpha^2 + \beta^2}}s\right)\right)\right)^2\right.$$

$$\left. + \left(\frac{c\alpha}{\sqrt{c^2\alpha^2 + \beta^2}}\sin\left(\alpha\left(a + \frac{1}{\sqrt{c^2\alpha^2 + \beta^2}}s\right)\right)\right)^2 + \left(\frac{\beta}{\sqrt{c^2\alpha^2 + \beta^2}}\right)^2\right)^{\frac{1}{2}}$$

$$= \sqrt{\frac{c^2\alpha^2}{c^2\alpha^2 + \beta^2} + \frac{\beta^2}{c^2\alpha^2 + \beta^2}}$$

$$= 1, \quad s \in [0, (b-a)\sqrt{c^2\alpha^2 + \beta^2}].$$

Thus, $g : [0, (b-a)\sqrt{c^2\alpha^2 + \beta^2}] \to \mathbb{R}^3$ is a naturally parameterized curve and the curves f and g are equivalent.

Exercise 1.7. Prove that the curves in Exercise 1.4 are not naturally parameterized.

1.2 Analytical representations of curves

Let $I \subset \mathbb{R}$.

Definition 1.7. A subset $M \subset \mathbb{R}^n$ is called a regular curve or a 1-dimensional smooth manifold of $\mathbb{R}^n$ if for each point $t_0 \in M$ there is a regular parameterized curve $f : I \to \mathbb{R}^n$ whose support $f(I)$ is an open neighborhood in M of the point t_0, i.e., is a set of the form $M \cap U$, where U is an neighborhood of t_0 in $\mathbb{R}^n$, while the map $f : I \to f(I)$ is a homeomorphism with respect to the topology of subspace of $f(I)$. A parameterized curve with these properties is called a local parametrization of the curve M around the point t_0. If for a curve M there is a local parametrization which is global, i.e., $f(I) = M$, the curve is called a simple curve.

1.2.1 Plane curves

Definition 1.8. A regular curve $M \subset \mathbb{R}^3$ is called a plane curve if it is contained in a plane Π. We shall usually assume that the plane Π coincides with the coordinate plane xOy. Here $O = (0, 0)$.

1.2.1.1 Parametric representation

We choose an arbitrary local parametrization $f(t) = (f_1(t), f_2(t))$ of a curve. Then the support $f(I)$ of this local parametrization is an open subset of the curve. For a global parametrization of a simple curve, $f(I)$ describes the entire curve. Thus, any point t_0 of the curve has an open neighborhood which is the support of the parameterized curve

$$x = f_1(t),$$
$$y = f_2(t).$$

(1.1)

Definition 1.9. The equations (1.1) are called the parametric equations of a curve in a neighborhood of a point t_0. Usually, unless the curve is simple, we cannot use the same set of equations to describe the points of an entire curve.

Example 1.14. Let $I = [0, \pi]$. Then

$$f_1(t) = t^2 + t + 1,$$
$$f_2(t) = 10 + \sin t + \cos t + \frac{1 - t + t^2}{1 + t^2 + t^4}, \quad t \in I,$$

is a parametric representation of a plane curve drawn in Fig. 1.8.

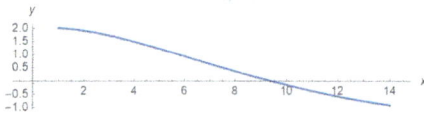

Figure 1.8: A parametric curve given by $f(t) = (t^2 + t + 1, 10 + \sin t + \cos t + (1 - t + t^2)/(1 + t^2 + t^4))$, $t \in [0, \pi]$.

Example 1.15. Let $I = [0, 2\pi]$. Then

$$f_1(t) = a(\cos t)^3,$$
$$f_2(t) = a(\sin t)^3, \quad t \in [0, 2\pi],$$

where a is a positive constant, is a parametric representation of the astroid. See Fig. 1.9 for the value $a = 1$.

Example 1.16. Let $I = [1, 2\pi]$. Then

$$f_1(t) = \frac{a}{2}\left(t + \frac{1}{t}\right),$$
$$f_2(t) = \frac{b}{2}\left(t - \frac{1}{t}\right), \quad t \in I,$$

where a and b are positive constants, is a parametric representation of a hyperbola. See Fig. 1.10 for the values of $a = b = 1$.

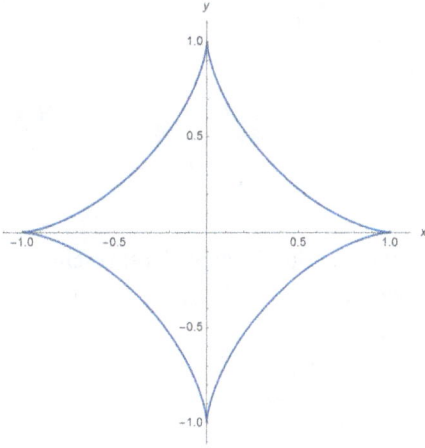

Figure 1.9: An astroid parametrically given by $f(t) = ((\cos t)^3, (\sin t)^3), t \in [0, 2\pi]$.

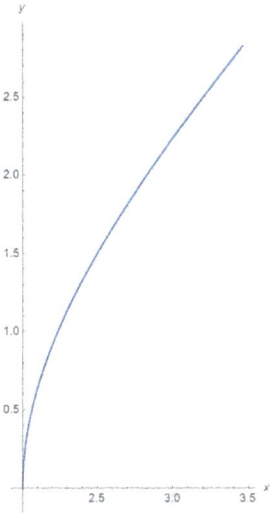

Figure 1.10: A hyperbola parametrically given by $f(t) = (t + 1/t, t - 1/t), t \in [0, 2\pi]$.

1.2.1.2 Explicit representation

Suppose that I is an open interval in $\mathbb{R}$ and $f : I \rightarrow \mathbb{R}$ is a smooth function, i. e., $f \in C^1(I)$. Then its graph

$$C = \{(t, f(t)) : t \in I\} \tag{1.2}$$

is a simple curve, which has the global representation

$$x = t,$$
$$y = f(t), \quad t \in I.$$

Definition 1.10. The equation

$$y = f(x)$$

is called an explicit equation of the curve (1.2). Sometimes for the explicit representation of a plane curve one also uses the term *nonparametric form*.

Example 1.17. Let $I = (0, \pi)$. Then

$$y = \frac{x + 1}{1 + x + x^2}, \quad x \in I,$$

is an explicit representation of a plane curve, see Fig. 1.11.

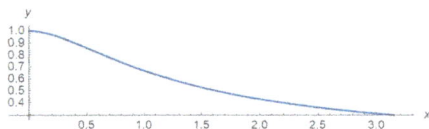

Figure 1.11: A curve with the explicit form $y = \frac{x+1}{1+x+x^2}, x \in [0, \pi]$.

Example 1.18. Let $I = (0, 15)$. Then

$$y = x^2, \quad x \in I,$$

is an explicit representation of a parabola.

Example 1.19. Let $I = (1, 24)$. Then

$$y = \frac{1}{1 + x}, \quad x \in I,$$

is an explicit equation of a hyperbola.

1.2.1.3 Implicit representation

Let $D \subset \mathbb{R}^2$. Let $F : D \to \mathbb{R}$ be a smooth function and suppose

$$C = \{(x, y) \in D : F(x, y) = 0\}$$

is the 0-level set of the function F. In the general case, C is not a regular curve. Nevertheless, if at the point $(x_0, y_0) \in C$ the gradient vector

$$\operatorname{grad} F(x_0, y_0) = (F'_x(x_0, y_0), F'_y(x_0, y_0))$$

is not vanishing, then there exists an open neighborhood U of the point (x_0, y_0) and a smooth function $y = f(x)$ defined on an open neighborhood $I \subset \mathbb{R}$ of the point x_0 such that

$$C \cap U = \{(x, f(x)) : x \in I\}.$$

If $\operatorname{grad} F \neq 0$ at all points of C, then C is a regular curve.

Example 1.20. The equation

$$(x^2 + y^2)^2 - 2b^2(x^2 - y^2) = a^4 - b^4,$$

where a and b are positive constants, is the equation of the Cassini oval. For the values $a = \sqrt{2}$ and $b = 1$, see Fig. 1.12.

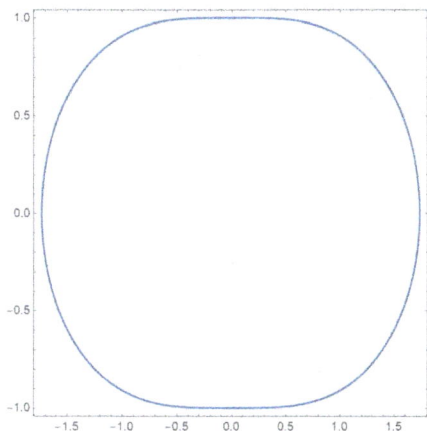

Figure 1.12: The Cassini oval where $a = \sqrt{2}$ and $b = 1$.

Example 1.21. The equation

$$(x^2 + y^2)^3 - 4a^2 x^2 y^2 = 0$$

is the equation of the four petal rosette. See Fig. 1.13 for $a = 1$.

Example 1.22. The equation

$$(x^2 + y^2 - 2ax)^2 = 4b^2(x^2 + y^2)$$

is the equation of the cardioid; see Fig. 1.14.

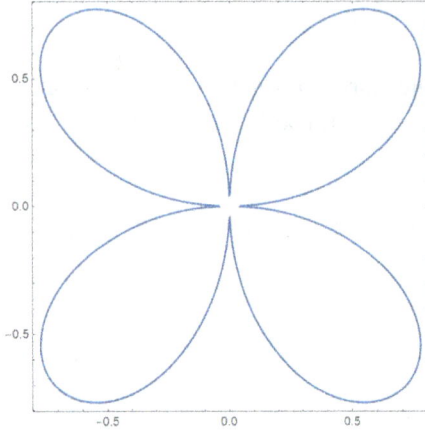

Figure 1.13: The four petal rosette where $a = 1$.

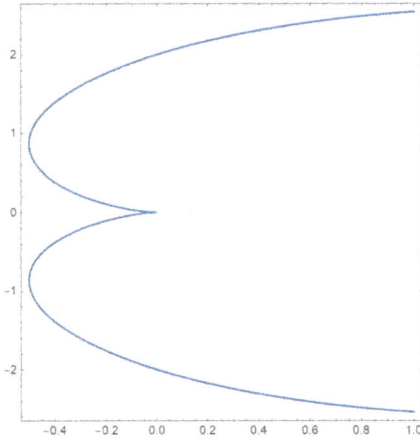

Figure 1.14: The cardioid where $a = 1/2$ and $b = 1/2$.

Remark 1.2. Note that the condition for nonsingularity of the grad F is only a sufficient condition for the equation $F(x, y) = 0$ to represent a curve. If grad $F(x_0, y_0) = 0$ for some $(x_0, y_0) \in D$, then we cannot claim that the equation represents a curve in a neighborhood of that point or the converse.

1.2.2 Space curves

1.2.2.1 Parametric representation
As in the case of plane curves, with local parametrization

$$x = f_1(t),$$
$$y = f_2(t),$$
$$z = f_3(t), \quad t \in I,$$

we can represent either the entire curve, or only a neighborhood of one of its points.

Example 1.23. Let $I = [0, \pi]$. Then

$$x = t + 1,$$
$$y = t^2 + 1,$$
$$z = \frac{1}{1 + t + t^2}, \quad t \in \mathbb{R},$$

is a parametric representation of a space curve; see Fig. 1.15.

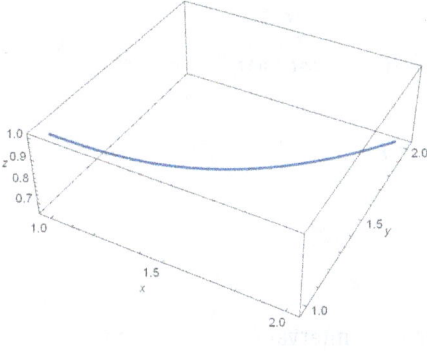

Figure 1.15: A parametrized space curve $f(t) = (t + 1, t^2 + t + 1, 1/(t^2 + t + 1))$, $t \in [0, 1]$.

Example 1.24. Let $I = [0, \pi]$. The equations

$$x = at \cos t,$$
$$y = at \sin t,$$
$$z = \frac{a^2 t^2}{2p}, \quad t \in I,$$

where a and p are positive constants, are a parametric representation of the Archimedes spiral. See Fig. 1.16 where $a = 1$ and $p = 2$.

Example 1.25. Let $I = \mathbb{R}$. The equations

$$x = \cos(3t),$$
$$y = \sin(3t),$$

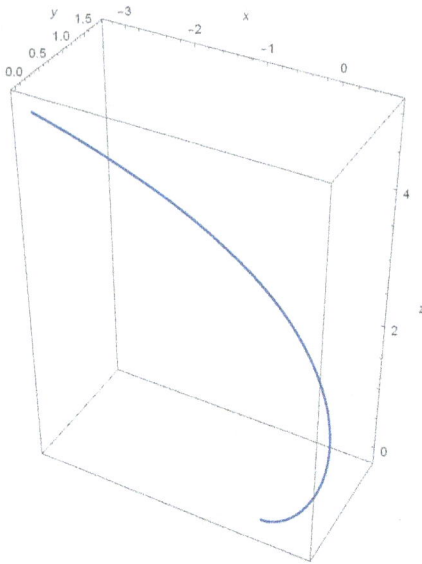

Figure 1.16: The Archimedes spiral parametrically given by $f(t) = (t\cos t, t\sin t, \frac{t^2}{2}), t \in [0, \pi]$.

$$z = 2t, \quad t \in I,$$

are a parametric representation of a circular helix.

1.2.2.2 Explicit representation

Let $f, g : I \to \mathbb{R}$ be two smooth functions on an open interval $I \subset \mathbb{R}$. Then the set

$$C = \{(x, f(x), g(x)) \subset \mathbb{R}^3 : x \in I\}$$

is a simple curve with a global representation given by

$$x = t,$$
$$y = f(t),$$
$$z = g(t), \quad t \in I.$$

Definition 1.11. The equations

$$y = f(x),$$
$$z = g(x), \quad x \in I,$$

are called the explicit equations of a space curve.

Example 1.26. Let $I = [1, 20]$. Then

$$y = x + 1$$
$$z = x^2 + x + 1, \quad x \in I,$$

is an explicit representation of a space curve.

Example 1.27. Let $I = [0, 2\pi]$. Then

$$x = \cos(4z),$$
$$y = \sin(4z),$$
$$z = z, \quad z \in I,$$

is an explicit representation of a circular helix.

1.2.2.3 Implicit representation

Let $D \subset \mathbb{R}^3$ and $F, G : D \to \mathbb{R}$ be smooth functions. Consider the set

$$C = \{(x, y, z) \in D : F(x, y, z) = 0, \quad G(x, y, z) = 0\},$$

i. e., the set of solutions of the system

$$F(x, y, z) = 0,$$
$$G(x, y, z) = 0.$$

In the general case, the set C is not a regular curve. Nevertheless, if, for some $a = (x_0, y_0, z_0) \in C$,

$$\text{rank} \begin{pmatrix} F'_x(a) & F'_y(a) & F'_z(a) \\ G'_x(a) & G'_y(a) & G'_z(a) \end{pmatrix} = 2,$$

then there is an open neighborhood $U \subset D$ of the point a such that $C \cap U$ is a curve. If the rank of the matrix

$$\begin{pmatrix} F'_x & F'_y & F'_z \\ G'_x & G'_y & G'_z \end{pmatrix}$$

is equal to two, then C is a regular curve.

Example 1.28. The equations

$$x^2 + y^2 + z^2 = 4b^2,$$
$$(x - b)^2 + y^2 = b^2,$$

where b is a positive constant, are the equations of the temple of Viviani.

1.3 The tangent and the normal plane

Suppose that $I \subseteq \mathbb{R}$, $t_0 \in I$, and $f : I \to \mathbb{R}^n$ is a curve so that $f'(t_0) \neq 0$.

Definition 1.12. The line passing through $f(t_0)$ and having direction of the vector $f'(t_0)$ is called the tangent of the curve at the point $f(t_0)$ (or at the point t_0).

The equations of the tangent line read as follows:

$$
\begin{aligned}
F(\lambda) &= f(t_0) + \lambda f'(t_0) \\
&= (f_1(t_0), f_2(t_0), \ldots, f_n(t_0)) + \lambda(f_1'(t_0), f_2'(t_0), \ldots, f_n'(t_0)) \\
&= (f_1(t_0), f_2(t_0), \ldots, f_n(t_0)) + (\lambda f_1'(t_0), \lambda f_2'(t_0), \ldots, \lambda f_n'(t_0)) \\
&= (f_1(t_0) + \lambda f_1'(t_0), f_2(t_0) + \lambda f_2'(t_0), \ldots, f_n(t_0) + \lambda f_n'(t_0)),
\end{aligned}
$$

with $\lambda \in \mathbb{R}$.

Example 1.29. Let

$$
f(t) = (t^2, t^3 + t, t), \quad t \in \mathbb{R};
$$

see Fig. 1.17. Here

$$
\begin{aligned}
f_1(t) &= t^2, \\
f_2(t) &= t^3 + t, \\
f_3(t) &= t, \quad t \in \mathbb{R}.
\end{aligned}
$$

Hence,

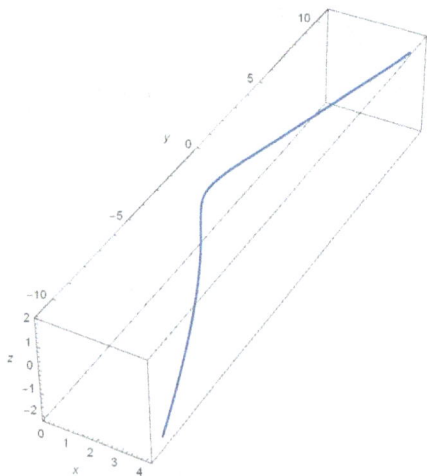

Figure 1.17: A parametric curve given by $f(t) = (t^2, t^3 + t, t), t \in [-2, 2]$.

$$f_1'(t) = 2t,$$
$$f_2'(t) = 3t^2 + 1,$$
$$f_3'(t) = 1, \quad t \in \mathbb{R},$$

and therefore the equation of the tangent line to f at an arbitrary point is given by

$$F(\lambda) = (t^2 + 2\lambda t, t^3 + t + \lambda(3t^2 + 1), t + \lambda), \quad \lambda, t \in \mathbb{R},$$

or

$$(3t^2 + 1)F_1(\lambda) + 2tF_2(\lambda) - 24(3t^3 + t)F_3(\lambda) + 7t^4 - 3t^2 = 0.$$

Example 1.30. Consider the curve

$$f(t) = (a \cosh t, a \sinh t, ct), \quad t \in \mathbb{R},$$

where $a, c \in \mathbb{R}, c \neq 0$. In the case $a = c = 1$, see Fig. 1.18. Here

$$f_1(t) = a \cosh t,$$
$$f_2(t) = a \sinh t,$$
$$f_3(t) = ct, \quad t \in I.$$

Hence,

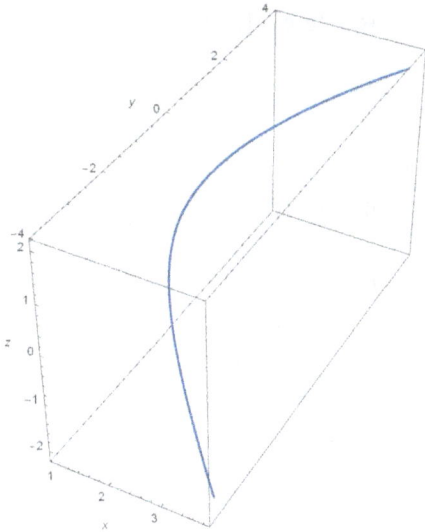

Figure 1.18: A parametric curve given by $f(t) = (\cosh t, \sinh t, t), t \in [-2, 2]$.

$$f_1'(t) = a \sinh t,$$
$$f_2'(t) = a \cosh t,$$
$$f_3'(t) = c, \quad t \in \mathbb{R}.$$

Then the equation of the tangent line at an arbitrary point is as follows:

$$(F_1(\lambda), F_2(\lambda), F_3(\lambda)) = (a \cosh t + \lambda a \sinh t, a \sinh t + \lambda a \cosh t, ct + c\lambda), \quad t, \lambda \in \mathbb{R},$$

or

$$\frac{F_1(\lambda) - a \cosh t}{a \sinh t} = \frac{F_2(\lambda) - a \sinh t}{a \cosh t} = \frac{F_3(\lambda) - ct}{c}, \quad \lambda, t \in \mathbb{R}.$$

Exercise 1.8. Find the equations of the tangent line at the corresponding points of the following curves:

1.
$$f(t) = (t, t^2 + 4t + 3), \quad t \in \mathbb{R}, \quad (-1, 0), (0, 3), (1, 8).$$

2.
$$f(t) = (t, t^3), \quad t \in \mathbb{R}, \quad (0, 0), (1, 1).$$

3.
$$f(t) = (t, \sin t), \quad t \in \mathbb{R}, \quad (0, 0), \left(\frac{\pi}{2}, 1\right), (\pi, 0).$$

4.
$$f(t) = (t, \tan t), \quad t \in \mathbb{R}, \quad (0, 0), \left(\frac{\pi}{4}, 1\right).$$

5.
$$f(t) = (e^t, e^{-t}, t^2), \quad t \in \mathbb{R}, \quad (e, e^{-1}, 1).$$

Answer 1.3. 1.

$$2F_1(\lambda) - F_2(\lambda) + 2 = 0,$$
$$4F_1(\lambda) - F_2(\lambda) + 3 = 0,$$
$$6F_1(\lambda) - F_2(\lambda) + 2 = 0, \quad \lambda \in \mathbb{R}.$$

2.

$$F_2(\lambda) = 0,$$
$$3F_1(\lambda) - F_2(\lambda) - 2 = 0, \quad \lambda \in \mathbb{R}.$$

3.

$$F_1(\lambda) = F_2(\lambda),$$
$$F_2(\lambda) = 1,$$
$$F_1(\lambda) + F_2(\lambda) = \pi, \quad \lambda \in \mathbb{R}.$$

4.

$$F_1(\lambda) = F_2(\lambda),$$

$$2F_1(\lambda) - F_2(\lambda) = \frac{\pi}{2} - 1, \quad \lambda \in \mathbb{R}.$$

5.

$$\frac{F_1(\lambda) - e}{e} = \frac{F_2(\lambda) - e^{-1}}{-e^{-1}} = \frac{F_3(\lambda) - 1}{2}, \quad \lambda \in \mathbb{R}.$$

Exercise 1.9. Prove that the tangent vectors of two equivalent curves are collinear at corresponding points and the tangent lines coincide.

Solution. Let (I, f) and (J, g) be two curves that are equivalent. Let $s : I \to J$ be the parameter change. Then

$$f(t) = g(s(t)), \quad t \in I.$$

Hence,

$$f'(t) = g'(s(t))s'(t), \quad t \in I.$$

This completes the proof.

Now, suppose that $f : I \to \mathbb{R}^n$ is a regular curve. For h, close enough to 0, or $h \to 0$, by the Taylor formula, we have

$$f(t_0 + h) = f(t_0) + hf'(t_0) + h\varepsilon,$$

where $\varepsilon \to 0$ as $h \to 0$. Let l be an arbitrary line passing through $f(t_0)$ and having unit direction vector m. Set

$$d(h) = d(f(t_0 + h), l),$$

where $d(f(t_0 + h), l)$ is the distance between $f(t_0 + h)$ and l.

Exercise 1.10. Prove that the line l is the tangent line to the regular parameterized curve f at the point t_0 if and only if

$$\lim_{h \to 0} \frac{d(h)}{|h|} = 0.$$

Solution. We have

$$
\begin{aligned}
d(h) &= |(f(t_0 + h) - f(t_0)) \times m| \\
&= |(hf'(t_0) + h\varepsilon) \times m| \\
&= |h(f'(t_0) \times m) + h(\varepsilon \times m)| \\
&= |h||(f'(t_0) \times m) + (\varepsilon \times m)|
\end{aligned}
$$

and

$$\frac{d(h)}{|h|} = |(f'(t_0) \times m) + (\varepsilon \times m)|.$$

Hence, using that $\varepsilon \to 0$ as $h \to 0$, we find

$$\lim_{h \to 0} \frac{d(h)}{|h|} = \lim_{h \to 0} |(f'(t_0) \times m) + (\varepsilon \times m)|$$
$$= |f'(t_0) \times m|.$$

1. Let l be the tangent line to f at t_0. Then $f'(t_0)$ and m are collinear and

$$|f'(t_0) \times m| = 0.$$

Hence,

$$\lim_{h \to 0} \frac{d(h)}{|h|} = 0. \tag{1.3}$$

2. Now, suppose that (1.3) holds. Then

$$|f'(t_0) \times m| = 0$$

and $f'(t_0)$ and m are collinear. Thus, l is a tangent line to f at t_0. This completes the proof.

Definition 1.13. Let (I, f) be a parametric curve in $\mathbb{R}^n$ ($n \geq 3$) and $t_0 \in I$. The normal (hyper)plane at $f(t_0)$ is the (hyper)plane through $f(t_0)$ that is perpendicular to the tangent line to the curve at the point $f(t_0)$.

The equation for the normal (hyper)plane is as follows:

$$(X - f(t_0)) \cdot f'(t_0) = 0, \quad t_0 \in I,$$

where $X = (x_1, \ldots, x_n)$ is an arbitrary point on the (hyper)plane and $\cdot$ denotes the inner product. We have

$$0 = ((x_1, \ldots, x_n) - (f_1(t_0), \ldots, f_n(t_0))) \cdot (f_1'(t_0), \ldots, f_n'(t_0))$$
$$= (x_1 - f_1(t_0), \ldots, x_n - f_n(t_0)) \cdot (f_1'(t_0), \ldots, f_n'(t_0))$$
$$= (x_1 - f_1(t_0))f_1'(t_0) + \cdots + (x_n - f_n(t_0))f_n'(t_0).$$

Example 1.31. Let $f : \mathbb{R} \to \mathbb{R}^3$ be defined by

$$f(t) = (t^2, e^{-t^2+1}, e^{-t+2}), \quad t \in \mathbb{R}.$$

Here

$$f_1(t) = t^2,$$
$$f_2(t) = e^{-t^2+1},$$
$$f_3(t) = e^{-t+2}, \quad t \in \mathbb{R}.$$

Hence,

$$f_1'(t) = 2t,$$
$$f_2'(t) = -2te^{-t^2+1},$$
$$f_3'(t) = -e^{-t+2}, \quad t \in \mathbb{R}.$$

Then the equation of the normal plane at an arbitrary point $t_0 \in \mathbb{R}$ is as follows:

$$2t_0(x_1 - t_0^2) - 2t_0 e^{-t_0^2+1}(x_2 - e^{-t_0^2+1}) - e^{t_0+2}(x_3 - e^{-t_0+2}) = 0.$$

Example 1.32. Let $f : \mathbb{R} \to \mathbb{R}^4$ be defined by

$$f(t) = (t, e^{-t}, e^{-t^2}, e^{-t^3}), \quad t \in \mathbb{R}.$$

Here

$$f_1(t) = t,$$
$$f_2(t) = e^{-t},$$
$$f_3(t) = e^{-t^2},$$
$$f_4(t) = e^{-t^3}, \quad t \in \mathbb{R}.$$

Hence,

$$f_1'(t) = 1,$$
$$f_2'(t) = -e^{-t},$$
$$f_3'(t) = -2te^{-t^2},$$
$$f_4'(t) = -3t^2 e^{-t^3}, \quad t \in \mathbb{R},$$

and the equation of the normal hyperplane at an arbitrary point $t_0 \in \mathbb{R}$ is as follows:

$$(x_1 - t_0) - e^{-t_0}(x_2 - e^{-t_0}) - 2t_0 e^{-t_0^2}(x_3 - e^{-t_0^2}) - 3t_0^2 e^{-t_0^3}(x_4 - e^{-t_0^3}) = 0.$$

Exercise 1.11. Find the equation of the normal plane at $t = 0$ for the following curve:

$$f(t) = (2\cos t, 2\sin t, 4t), \quad t \in \mathbb{R}.$$

Answer 1.4.

$$x_2 + x_3 = 0.$$

1.4 Osculating plane

Definition 1.14. A parameterized curve $f = f(t)$, $t \in I$, is said to be biregular at the point t_0 if the vectors $f'(t_0)$ and $f''(t_0)$ are not collinear, i. e., if

$$f'(t_0) \times f''(t_0) \neq 0.$$

Example 1.33. Let $f : \mathbb{R} \to \mathbb{R}^3$ be given by

$$f(t) = (t^2, e^t, e^{2t}), \quad t \in \mathbb{R};$$

see Fig. 1.19. Here

$$f_1(t) = t^2,$$
$$f_2(t) = e^t,$$
$$f_3(t) = e^{2t}, \quad t \in \mathbb{R}.$$

Hence,

$$f_1'(t) = 2t,$$

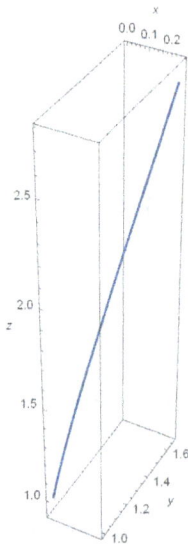

Figure 1.19: A parametric curve given by $f(t) = (t^2, e^t, e^{2t})$, $t \in [0, 1/2]$.

$$f_2'(t) = e^t,$$
$$f_3'(t) = 2e^{2t}, \quad t \in \mathbb{R},$$

and

$$f_1''(t) = 2,$$
$$f_2''(t) = e^t,$$
$$f_3''(t) = 4e^{2t}, \quad t \in \mathbb{R}.$$

Therefore

$$f'(t) = (2t, e^t, 2e^{2t}),$$
$$f''(t) = (2, e^t, 4e^{2t}), \quad t \in \mathbb{R},$$

and

$$f'(t) \times f''(t) = (2t, e^t, 2e^{2t}) \times (2, e^t, 4e^{2t})$$
$$= (4e^{3t} - 2e^{3t}, 4e^{2t} - 8te^{2t}, 2te^t - 2e^t)$$
$$= (2e^{3t}, (4 - 8t)e^{2t}, 2(t - 1)e^t)$$
$$\neq 0, \quad t \in \mathbb{R}.$$

Thus, the considered curve is a biregular curve.

Example 1.34. Consider the curve

$$f(t) = (t, t^2), \quad t \in [2, 4].$$

We have

$$f_1(t) = t,$$
$$f_2(t) = t^2, \quad t \in [2, 4].$$

Then

$$f_1'(t) = 1,$$
$$f_2'(t) = 2t, \quad t \in [2, 4],$$

and

$$f_1''(t) = 0,$$
$$f_2''(t) = 2, \quad t \in [2, 4].$$

Thus,

$$f'(t) = (f_1'(t), f_2'(t))$$
$$= (1, 2t), \quad t \in [2, 4],$$

and

$$f''(t) = (f_1''(t), f_2''(t))$$
$$= (0, 2), \quad t \in [2, 4].$$

Note that $f'(t)$ and $f''(t)$ are not collinear for any $t \in [2, 4]$ and thus the considered curve is biregular.

Exercise 1.12. Prove that the curve

$$f(t) = (t + 1, t^2 + t + 1, t^3), \quad t \in [2, 4],$$

is a biregular curve.

Definition 1.15. Let (I, f) be a parameterized curve in $\mathbb{R}^3$ and $t_0 \in I$. The osculating plane through $f(t_0)$ that is parallel to $f'(t_0)$ and $f''(t_0)$ is defined by

$$0 = ((x, y, z) - f(t_0)) \cdot (f'(t_0) \times f''(t_0)), \quad t_0 \in I.$$

Here we have that

$$0 = \begin{vmatrix} x - f_1(t_0) & y - f_2(t_0) & z - f_3(t_0) \\ f_1'(t_0) & f_2'(t_0) & f_3'(t_0) \\ f_1''(t_0) & f_2''(t_0) & f_3''(t_0) \end{vmatrix}, \quad t_0 \in I,$$

is the equation of the osculating plane.

Example 1.35. Let $f : \mathbb{R} \to \mathbb{R}^3$ be given by

$$f(t) = (t, t^2, t + t^2), \quad t \in \mathbb{R};$$

see Fig. 1.20. Here

$$f_1(t) = t,$$
$$f_2(t) = t^2,$$
$$f_3(t) = t + t^2, \quad t \in \mathbb{R}.$$

Then

$$f_1'(t) = 1,$$

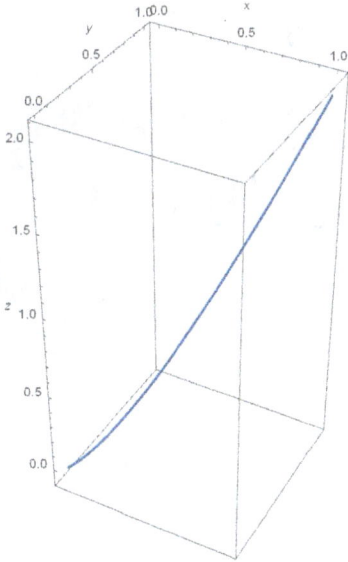

Figure 1.20: A parametric curve given by $f(t) = (t, t^2, t + t^2), t \in [0, 1/2]$.

$$f_2'(t) = 2t,$$
$$f_3'(t) = 1 + 2t, \quad t \in \mathbb{R},$$

and

$$f_1''(t) = 0,$$
$$f_2''(t) = 2,$$
$$f_3''(t) = 2, \quad t \in \mathbb{R}.$$

Therefore,

$$f'(t) = (f_1'(t), f_2'(t), f_3'(t))$$
$$= (1, 2t, 1 + 2t),$$
$$f''(t) = (f_1''(t), f_2''(t), f_3''(t))$$
$$= (0, 2, 2), \quad t \in \mathbb{R},$$

and the equation of the osculating plane is

$$0 = \begin{vmatrix} x - t_0 & y - t_0^2 & z - t_0 - t_0^2 \\ 1 & 2t_0 & 1 + 2t_0 \\ 0 & 2 & 2 \end{vmatrix}$$
$$= -2(x - t_0) - 2(y - t_0^2) + 2(z - t_0 - t_0^2)$$
$$= -2x - 2y + 2z + 4t_0^2, \quad t_0 \in \mathbb{R},$$

or

$$x + y - z - 4t_0^2 = 0, \quad t_0 \in \mathbb{R}.$$

Example 1.36. Let $f : \mathbb{R} \rightarrow \mathbb{R}^3$ be given by

$$f(t) = (e^t, e^{t^2}, e^{t^3}), \quad t \in \mathbb{R};$$

see Fig. 1.21. Here

$$f_1(t) = e^t,$$
$$f_2(t) = e^{t^2},$$
$$f_3(t) = e^{t^3}, \quad t \in \mathbb{R}.$$

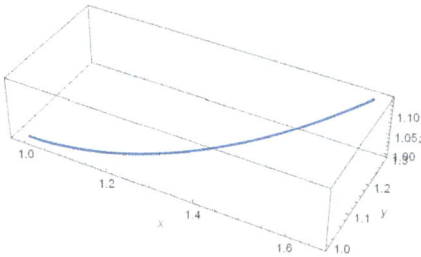

Figure 1.21: A parametric curve given by $f(t) = (e^t, e^{t^2}, e^{t^3})$, $t \in [0, 1/2]$.

Hence,

$$f_1'(t) = e^t,$$
$$f_2'(t) = 2te^{t^2},$$
$$f_3'(t) = 3t^2 e^{t^3}, \quad t \in \mathbb{R},$$

and

$$f_1''(t) = e^t,$$
$$f_2''(t) = 2e^{t^2} + 4t^2 e^{t^2},$$
$$f_3''(t) = 6te^{t^3} + 9t^4 e^{t^3}, \quad t \in \mathbb{R}.$$

Therefore the equation of the osculating plane at $t_0 = 0$ is

$$0 = \begin{vmatrix} x-1 & y-1 & z-1 \\ 1 & 0 & 0 \\ 1 & 1 & 0 \end{vmatrix}$$

$$= z - 1,$$

or

$$z = 1.$$

Exercise 1.13. Find the osculating plane at an arbitrary point $t_0 \in \mathbb{R}$ for the following curve:

$$f(t) = (\cos t, \sin t, t), \quad t \in \mathbb{R}.$$

Answer 1.5.

$$\sin t_0 x - \cos t_0 y + z - t_0 = 0, \quad t_0 \in \mathbb{R}.$$

Exercise 1.14. Prove that the osculating planes of two equivalent curves coincide at the biregular points.

Solution. Let (I, f) and (J, g) be two equivalent curves. Let also $s : I \rightarrow J$ be the parameter change. Then

$$f(t) = g(s(t)),$$
$$f'(t) = g'(s(t)) \cdot s'(t),$$
$$f''(t) = f''(t) \cdot (s'(t))^2 + g'(s(t)) \cdot s''(t), \quad t \in I.$$

Thus, the sets $\{f'(t), f''(t)\}$ and $\{g'(s(t)), g''(s(t))\}$ are linearly dependent, namely these two sets of vectors describe a unique plane. This completes the proof.

Let (I, f) be a parameterized curve in $\mathbb{R}^3$, $t_0 \in I$, and $f(t_0)$ be biregular. Suppose that α is a plane with a unit normal vector e and α passes through $f(t_0)$. Denote

$$d(h) = d(f(t_0 + h), \alpha).$$

Exercise 1.15. Prove that the plane α is an osculating plane to the curve $f = f(t)$ at the biregular point $f(t_0)$ if and only if

$$\lim_{h \to 0} d(h)/|h|^2 = 0. \tag{1.4}$$

Solution. By the Taylor formula, we have

$$f(t_0 + h) = f(t_0) + hf'(t_0) + \frac{1}{2}h^2 \cdot f''(t_0) + h^2 \varepsilon,$$

where $\varepsilon \to 0$ as $h \to 0$. From here,

$$d(h) = \left| e \cdot (f(t_0 + h) - f(t_0)) \right|$$
$$= \left| e \cdot \left(hf'(t_0) + \frac{1}{2}h^2 f''(t_0) + h^2 \varepsilon \right) \right|$$
$$= \left| e \cdot (hf'(t_0)) + e \cdot \left(\frac{1}{2}h^2 f''(t_0) \right) + e \cdot (h^2 \varepsilon) \right|,$$

thus

$$\frac{d(h)}{h^2} = \left| \frac{1}{h}(e \cdot f'(t_0)) + \frac{1}{2}(e \cdot f''(t_0)) + e \cdot \varepsilon \right|$$

and

$$\lim_{h \to 0} \frac{d(h)}{h^2} = \lim_{h \to 0} \left| \frac{1}{h}(e \cdot f'(t_0)) + \frac{1}{2}(e \cdot f''(t_0)) + e \cdot \varepsilon \right|. \tag{1.5}$$

1. Let (1.4) hold. Then

$$e \cdot f'(t_0) = 0,$$
$$e \cdot f''(t_0) = 0. \tag{1.6}$$

Thus,

$$e \parallel f'(t_0) \times f''(t_0)$$

and α is an osculating plane.
2. Assume α is an osculating plane. Then, we have (1.6) and, using (1.5), we get (1.4). This completes the proof.

1.5 Curvature of a curve

Let (I, f) and (J, g) be two curves in $\mathbb{R}^n$ that are equivalent with the parameter change s. Then

$$f(t) = g(s(t)), \quad t \in I.$$

We have $s : I \to J, s' > 0$ on I, and

$$|g'(s)| = 1.$$

Exercise 1.16. Prove that the vector $g''(s)$ in $\mathbb{R}^n$ does not depend on the parameter change.

Solution. Let (J_1, g_1) be another naturally parameterized curve with the parameter change s_1, i. e.,

$$g(s) = g_1(s_1(s)), \quad s \in J,$$

where $s_1 : J \to J_1$. Then

$$g'(s) = g_1'(s_1(s))s_1'(s), \quad s \in J.$$

Hence,

$$|s_1'(s)| = 1$$

and

$$s_1'(s) = \pm 1, \quad s \in J.$$

Therefore

$$s_1(s) = \pm s + s_0,$$

for some constant $s_0 \in \mathbb{R}$. Then

$$s_1''(s) = 0, \quad s \in J,$$

and

$$\begin{aligned} g''(s) &= g_1''(s_1(s))(s_1'(s))^2 + g'(s_1(s))s_1''(s) \\ &= g_1''(s_1(s))(s_1'(s))^2 \\ &= g_1''(s_1(s)), \quad s \in J. \end{aligned}$$

This completes the proof.

Definition 1.16. The vector

$$\mathbf{k}(t) = g''(s(t))$$

is called the curvature vector of the curve $f = f(t)$ in $\mathbb{R}^n$ at the point t and

$$\kappa(t) = |g''(s(t))|$$

is called the curvature of f at the point t.

Assume that $n = 3$. We have

$$f'(t) = g'(s(t))s'(t), \quad t \in I,$$

and

$$s'(t) = |f'(t)|, \quad t \in I.$$

Therefore,

$$g'(s(t)) = \frac{1}{s'(t)} f'(t)$$
$$= \frac{1}{|f'(t)|} f'(t), \quad t \in I.$$

Furthermore,

$$s''(t) = |f'|'(t)$$
$$= \frac{d((f'(t) \cdot f'(t))^{\frac{1}{2}})}{dt}$$
$$= \frac{f'(t) \cdot f''(t)}{|f'(t)|}, \quad t \in I.$$

Therefore,

$$f''(t) = g''(s(t))(s'(t))^2 + g'(s(t))s''(t), \quad t \in I,$$

from where

$$g''(s(t))(s'(t))^2 = f''(t) - \frac{(f'(t) \cdot f''(t))}{|f'(t)|^2} f'(t), \quad t \in I,$$

or

$$g''(s(t))|f'(t)|^2 = f''(t) - \frac{(f'(t) \cdot f''(t))}{|f'(t)|^2} f'(t), \quad t \in I,$$

or

$$g''(s(t)) = \frac{f''(t)}{|f'(t)|^2} - \frac{(f'(t) \cdot f''(t))f'(t)}{|f'(t)|^4}, \quad t \in I.$$

Consequently,

$$\mathbf{k}(t) = \frac{f''(t)}{|f'(t)|^2} - \frac{(f'(t) \cdot f''(t))f'(t)}{|f'(t)|^4}, \quad t \in I,$$

and

$$\kappa(t) = |\mathbf{k}(t)|$$

$$= \left| \frac{f''(t)}{|f'(t)|^2} - \frac{(f'(t) \cdot f''(t))f'(t)}{|f'(t)|^4} \right|$$

$$= \frac{1}{|f'(t)|^4} \left| |f''(t)||f'(t)|^2 - (f'(t) \cdot f''(t))f'(t) \right|$$

$$= \frac{1}{|f'(t)|^3} \sqrt{(f''(t) \cdot f''(t))(f'(t) \cdot f'(t)) - (f'(t) \cdot f''(t))^2}$$

$$= \frac{|f'(t) \times f''(t)|}{|f'(t)|^3}, \quad t \in I.$$

This formula is also valid for $n = 2$.

Example 1.37. Consider the curve

$$f(t) = (t, \sin t), \quad t \in \mathbb{R}.$$

We will find its curvature. Here

$$f_1(t) = t,$$
$$f_2(t) = \sin t, \quad t \in \mathbb{R}.$$

Then

$$f_1'(t) = 1,$$
$$f_2'(t) = \cos t, \quad t \in \mathbb{R},$$

and

$$f_1''(t) = 0,$$
$$f_2''(t) = -\sin t, \quad t \in \mathbb{R}.$$

Hence,

$$f'(t) = (f_1'(t), f_2'(t))$$
$$= (1, \cos t), \quad t \in \mathbb{R},$$

and

$$f''(t) = (f_1''(t), f_2''(t))$$
$$= (0, -\sin t), \quad t \in \mathbb{R}.$$

Then

$$|f'(t) \times f''(t)| = |\sin t|,$$
$$|f'(t)| = \left((f_1'(t))^2 + (f_2'(t))^2\right)^{\frac{1}{2}}$$
$$= \left(1 + (\cos t)^2\right)^{\frac{1}{2}}, \quad t \in \mathbb{R}.$$

Thus

$$\kappa(t) = \frac{|\sin t|}{(1 + (\cos t)^2)^{\frac{3}{2}}}, \quad t \in \mathbb{R}.$$

Example 1.38. We will find the curvature of the following curve:

$$f(t) = (a \cosh t, a \sinh t, at), \quad t \in \mathbb{R},$$

where $a \in \mathbb{R}$ is a parameter, $a \neq 0$. Here

$$f_1(t) = a \cosh t,$$
$$f_2(t) = a \sinh t,$$
$$f_3(t) = at, \quad t \in \mathbb{R}.$$

Then

$$f_1'(t) = a \sinh t,$$
$$f_2'(t) = a \cosh t,$$
$$f_3'(t) = a, \quad t \in \mathbb{R},$$

and

$$f_1''(t) = a \cosh t,$$
$$f_2''(t) = a \sinh t,$$
$$f_3''(t) = 0, \quad t \in \mathbb{R}.$$

Hence,

$$f'(t) = (f_1'(t), f_2'(t), f_3'(t))$$
$$= (a \sinh t, a \cosh t, a),$$
$$f''(t) = (f_1''(t), f_2''(t), f_3''(t))$$
$$= (a \cosh t, a \sinh t, 0), \quad t \in \mathbb{R},$$
$$f'(t) \times f''(t) = (-a^2 \sinh t, a^2 \cosh t, -a^2), \quad t \in \mathbb{R},$$

and

$$|f'(t) \times f''(t)| = \sqrt{\left(-a^2 \sinh t\right)^2 + \left(a^2 \cosh t\right)^2 + \left(-a^2\right)^2}$$
$$= a^2 \sqrt{1 + (\sinh t)^2 + (\cosh t)^2}$$
$$= a^2 \sqrt{1 + \cosh(2t)},$$
$$|f'(t)| = \sqrt{(a \sinh t)^2 + (a \cosh t)^2 + a^2}$$
$$= |a| \sqrt{1 + (\sinh t)^2 + (\cosh t)^2}$$
$$= |a| \sqrt{1 + \cosh(2t)}, \quad t \in \mathbb{R}.$$

Therefore

$$\kappa(t) = \frac{|f'(t) \times f''(t)|}{|f'(t)|^3}$$
$$= \frac{a^2 \sqrt{1 + \cosh(2t)}}{|a|^3 (1 + \cosh(2t))^{\frac{3}{2}}}$$
$$= \frac{1}{|a|(1 + \cosh(2t))}, \quad t \in \mathbb{R}.$$

Exercise 1.17. Find the curvature of the following curves:

1.
$$f(t) = (t^2, t^3), \quad t \in \mathbb{R}.$$

2.
$$f(t) = (a \cos t, b \sin t), \quad t \in \mathbb{R},$$

where $a, b \in \mathbb{R}$ such that $a^2 + b^2 \neq 0$.

3.
$$f(t) = (a \cosh t, b \sinh t), \quad t \in \mathbb{R},$$

where $a, b \in \mathbb{R}$ such that $a^2 + b^2 \neq 0$.

4.
$$f(t) = (a \cos t, a \sin t, bt), \quad t \in \mathbb{R},$$

where $a, b \in \mathbb{R}$ such that $a^2 + b^2 \neq 0$.

5.
$$f(t) = (t \cos t, t \sin t, at), \quad t \in \mathbb{R},$$

where $a \in \mathbb{R}$ is a parameter, $a \neq 0$.

Answer 1.6. 1.
$$\kappa(t) = \frac{6}{t(4 + 9t^2)^{\frac{3}{2}}}, \quad t \in \mathbb{R}.$$

2.
$$\kappa(t) = \frac{ab}{(a^2(\sin t)^2 + b^2(\cos t)^2)^{\frac{3}{2}}}, \quad t \in \mathbb{R}.$$

3.

$$\kappa(t) = \frac{ab}{(a^2(\sinh t)^2 + b^2(\cosh t)^2)^{\frac{3}{2}}}, \quad t \in \mathbb{R}.$$

4.

$$\kappa(t) = \frac{|a|}{a^2 + b^2}, \quad t \in \mathbb{R}.$$

5.

$$\kappa(t) = \frac{\sqrt{2 + t^2}}{2 + a^2 + t^2}, \quad t \in \mathbb{R}.$$

1.6 The Frenet frame, Frenet formulae, and the torsion

Suppose that $(I, f), f : \mathbb{R} \to \mathbb{R}^3$ is a biregular curve.

Definition 1.17. The Frenet frame or the moving frame at the point t_0 is an orthonormal frame in $\mathbb{R}^3$ with the origin at the point $f(t_0)$ and the coordinate vectors $\{\mathbf{t}(t_0), \mathbf{n}(t_0), \mathbf{b}(t_0)\}$, where

1.

$$\mathbf{t}(t_0) = \frac{1}{|f'(t_0)|} f'(t_0),$$

which is called the unit tangent at t_0;

2.

$$\mathbf{n}(t_0) = \frac{1}{|k(t_0)|} \mathbf{k}(t_0),$$

which is called the unit principal normal at t_0;

3.

$$\mathbf{b}(t_0) = \mathbf{t}(t_0) \times \mathbf{n}(t_0),$$

which is called the binormal at t_0.

For a naturally parameterized curve (J, g), we have

$$\mathbf{t}(s_0) = g'(s_0),$$

$$\mathbf{n}(s_0) = \frac{1}{|g''(s_0)|} g''(s_0),$$

$$\mathbf{b}(s_0) = \frac{1}{|g''(s_0)|} g'(s_0) \times g''(s_0).$$

We have

$$\mathbf{t}(t) = \frac{1}{|f'(t)|} f'(t), \quad t \in I,$$

$$\mathbf{k}(t) = \frac{1}{|f'(t)|^2} f''(t) - \frac{f'(t) \cdot f''(t)}{|f'(t)|^4} f'(t), \quad t \in I,$$

and

$$k(t) = \frac{|f'(t) \times f''(t)|}{|f'(t)|^3}, \quad t \in I.$$

Hence,

$$\mathbf{n}(t) = \frac{1}{k(t)} \mathbf{k}(t)$$

$$= \frac{|f'(t)|}{|f' \times f''(t)|} f''(t) - \frac{f'(t) \cdot f''(t)}{|f'(t) \times f''(t)||f'(t)|} f'(t), \quad t \in I,$$

and

$$\mathbf{b}(t) = \mathbf{t}(t) \times \mathbf{n}(t)$$

$$= \frac{1}{|f'(t) \times f''(t)|} f'(t) \times f''(t), \quad t \in I$$

Next,

$$\mathbf{b}(t) \times \mathbf{t}(t) = (\mathbf{t}(t) \times \mathbf{n}(t)) \times \mathbf{t}(t)$$

$$= (\mathbf{t}(t) \cdot \mathbf{t}(t))\mathbf{n}(t) - (\mathbf{n}(t) \cdot \mathbf{t}(t))\mathbf{t}(t)$$

$$= \mathbf{n}(t), \quad t \in I.$$

Example 1.39. Consider the curve given in Example 1.38. Using the computations there, we get

$$\mathbf{t}(t) = \frac{1}{|a| \sqrt{1 + \cosh(2t)}} (a \sinh t, a \cosh t, a)$$

$$= \frac{\operatorname{sign}(a)}{\sqrt{1 + \cosh(2t)}} (\sinh t, \cosh t, 1), \quad t \in \mathbb{R},$$

and

$$f'(t) \cdot f''(t) = a^2 \sinh t \cosh t + a^2 \sinh t \cosh t$$

$$= 2a^2 \sinh t \cosh t$$

$$= a^2 \sinh(2t), \quad t \in \mathbb{R}.$$

Then

$$\mathbf{k}(t) = \frac{1}{a^2(1 + \cosh(2t))} (a \cosh t, a \sinh t, 0)$$

$$- \frac{a^2 \sinh(2t)}{a^4(1 + \cosh(2t))^2} (a \sinh t, a \cosh t, a)$$

$$= \frac{1}{a(1 + \cosh(2t))} (\cosh t, \sinh t, 0) - \frac{\sinh(2t)}{a(1 + \cosh(2t))^2} (\sinh t, \cosh t, 1)$$

$$= \frac{1}{a(1+\cosh(2t))^2}\big((1+\cosh(2t))\cosh t - \sinh(2t)\sinh t,$$

$$(1+\cosh(2t))\sinh t - \sinh(2t)\cosh t, -\sinh(2t)\big)$$

$$= \frac{1}{a(1+\cosh(2t))^2}(\cosh t + \cosh t, \sinh t - \sinh t, -\sinh(2t))$$

$$= \frac{1}{a(1+\cosh(2t))^2}(2\cosh t, 0, -\sinh(2t)), \quad t \in \mathbb{R},$$

$$\mathbf{n}(t) = |a|(1+\cosh(2t)) \cdot \frac{1}{a(1+\cosh(2t))^2}(2\cosh t, 0, -\sinh(2t))$$

$$= \frac{\operatorname{sign}(a)}{1+\cosh(2t)}(2\cosh t, 0, -\sinh(2t)), \quad t \in \mathbb{R},$$

and

$$\mathbf{b}(t) = \frac{1}{(1+\cosh(2t))^{\frac{3}{2}}}\big(-\cosh t \sinh(2t), 2\cosh t + \sinh t \sinh(2t), -2(\cosh t)^2\big)$$

$$= \frac{1}{(1+\cosh(2t))^{\frac{3}{2}}}\big(-\cosh t \sinh(2t), 2\cosh t(1+(\sinh t)^2), -2(\cosh t)^2\big)$$

$$= -\frac{\cosh t}{(1+\cosh(2t))^{\frac{3}{2}}}\big(\sinh(2t), 2(\cosh t)^2, -2\cosh t\big), \quad t \in \mathbb{R}.$$

Definition 1.18. An orientation of a regular curve $C \subset \mathbb{R}^3$ is a family of local parameterizations $\{(I_\alpha, f_\alpha)\}_{\alpha \in A}$ such that

1. $C = \bigcup_{\alpha \in A} f_\alpha(I_\alpha)$,
2.   for any connected component $C_{\alpha\beta}^b$ of

$$C_{\alpha\beta} = f_\alpha(I_\alpha) \cap f_\beta(I_\beta), \quad \alpha, \beta \in A,$$

the parameterized curves (I_α^b, f_α^b) and (I_β^b, f_β^b), with

$$I_\alpha^b = f_\alpha^{-1}(C_{\alpha\beta}^b),$$
$$f_\alpha^b = f_\alpha|_{I_\alpha^b},$$
$$I_\beta^b = f_\beta^{-1}(C_{\alpha\beta}^b),$$
$$f_\beta^b = f_\beta|_{I_\beta^b},$$

are positively equivalent.

Definition 1.19. A regular curve $C \subset \mathbb{R}^3$ with an orientation is called an oriented regular curve.

Definition 1.20. A local parameterization (I, f) of an oriented regular curve C is called compatible with the orientation defined by the family $\{(I_\alpha, f_\alpha)\}_{\alpha \in A}$ if on the intersections $f(I) \cap f_\alpha(I_\alpha)$ the parameterized curves (I, f) and (I_α, f_α) are positively oriented.

Definition 1.21. The Frenet frame of an oriented biregular curve C at a point $x \in C$ is the Frenet frame of a biregular parameterized curve $f = f(t)$ at t_0, where $f = f(t)$ is a local parameterization of the curve C, compatible with the orientation such that $f(t_0) = x$.

Let (I, f) be a biregular curve. Let also $\{\mathbf{t}, \mathbf{n}, \mathbf{b}\}$ be the Frenet frame. Then

$$\mathbf{t}(t) = \frac{f'(t)}{|f'(t)|}, \quad t \in I,$$

and

$$
\begin{aligned}
\mathbf{t}'(t) &= \frac{f''(t)|f'(t)| - f'(t)(\frac{f'(t) \cdot f''(t)}{|f'(t)|})}{|f'(t)|^2} \\
&= \frac{f''(t)|f'(t)|^2 - f'(t)(f'(t) \cdot f''(t))}{|f'(t)|^3} \\
&= |f'(t)|\left(\frac{|f'(t) \times f''(t)|}{|f'(t)|^3}\right)\left(\frac{|f'(t)|}{|f'(t) \times f''(t)|}f''(t) - \frac{f'(t) \cdot f''(t)}{|f'(t)||f'(t) \times f''(t)|}f'(t)\right) \\
&= |f'(t)|k(t)\mathbf{n}(t), \quad t \in I.
\end{aligned}
$$

Furthermore,

$$\mathbf{b}(t) = \mathbf{t}(t) \times \mathbf{n}(t), \quad t \in I,$$

whereupon

$$
\begin{aligned}
\mathbf{b}'(t) &= \mathbf{t}'(t) \times \mathbf{n}(t) + \mathbf{t}(t) \times \mathbf{n}'(t) \\
&= |f'(t)|k(t)(\mathbf{n}(t) \times \mathbf{n}(t)) + \mathbf{t}(t) \times \mathbf{n}'(t) \\
&= \mathbf{t}(t) \times \mathbf{n}'(t), \quad t \in I.
\end{aligned}
$$

Therefore

$$\mathbf{b}'(t) \perp \mathbf{t}(t), \quad t \in I,$$

and from the equality

$$|\mathbf{b}(t)|^2 = 1, \quad t \in I,$$

we get

$$\mathbf{b}'(t) \cdot \mathbf{b}(t) = 0, \quad t \in I,$$

and then

$$\mathbf{b}'(t) \perp \mathbf{b}(t), \quad t \in I.$$

Therefore, $\mathbf{b}'(t)$, $t \in I$, is collinear with

$$\mathbf{b}(t) \times \mathbf{t}(t), \quad t \in I,$$

or $\mathbf{b}'(t)$, $t \in I$, is collinear with $\mathbf{n}(t)$, $t \in I$. We can write

$$\mathbf{b}'(t) = -|f'(t)|\tau(t)\mathbf{n}(t), \quad t \in I.$$

From the equation

$$\mathbf{n}(t) = \mathbf{b}(t) \times \mathbf{t}(t), \quad t \in I,$$

we find

$$
\begin{aligned}
\mathbf{n}'(t) &= \mathbf{b}'(t) \times \mathbf{t}(t) + \mathbf{b}(t) \times \mathbf{t}'(t) \\
&= (-|f'(t)|\tau(t)\mathbf{n}(t)) \times \mathbf{t}(t) + \mathbf{b}(t) \times (|f'(t)|\kappa(t)\mathbf{n}(t)) \\
&= -|f'(t)|\tau(t)(\mathbf{n}(t) \times \mathbf{t}(t)) + |f'(t)|\kappa(t)(\mathbf{b}(t) \times \mathbf{n}(t)) \\
&= |f'(t)|\tau(t)\mathbf{b}(t) - |f'(t)|\tau(t)\mathbf{t}(t) \\
&= |f'(t)|(-\kappa(t)\mathbf{t}(t) + \tau(t)\mathbf{b}(t)), \quad t \in I.
\end{aligned}
$$

Therefore

$$
\begin{aligned}
\mathbf{t}'(t) &= |f'(t)|\kappa(t)\mathbf{n}(t), \\
\mathbf{b}'(t) &= -|f'(t)|\tau(t)\mathbf{n}(t), \\
\mathbf{n}'(t) &= |f'(t)|(-\kappa(t)\mathbf{t}(t) + \tau(t)\mathbf{b}(t)), \quad t \in I.
\end{aligned}
\tag{1.7}
$$

Definition 1.22. The formulae (1.7) is called Frenet formulae.

Definition 1.23. The quantity $\tau(t)$, $t \in I$, is called torsion or the second curvature.

Now, we will deal with a naturally parameterized curve $(J, g = g(s))$. Then the Frenet formulae take the form

$$
\begin{aligned}
\mathbf{t}'(s) &= \kappa(s)\mathbf{n}(s), \\
\mathbf{b}'(s) &= -\tau(s)\mathbf{n}(s), \\
\mathbf{n}'(s) &= -\kappa(s)\mathbf{t}(s) + \tau(s)\mathbf{b}(s), \quad s \in J.
\end{aligned}
$$

The Frenet frame is as follows:

$$
\begin{aligned}
\mathbf{t} &= g', \\
\mathbf{n} &= (1/\kappa)g'', \\
\mathbf{b} &= (1/\kappa)(g' \times g'').
\end{aligned}
\tag{1.8}
$$

Then

$$\mathbf{b}'(s) \cdot \mathbf{n}(s) = (-\tau(s)\mathbf{n}(s)) \cdot \mathbf{n}(s)$$
$$= -\tau(s)(\mathbf{n}(s) \cdot \mathbf{n}(s))$$
$$= -\tau(s), \quad s \in J.$$

Using the third equation of (1.8), we find

$$\mathbf{b}'(s) = \left(\frac{1}{\kappa(s)}\right)'(g'(s) \times g''(s)) + \frac{1}{\kappa(s)}(g''(s) \times g''(s)) + \frac{1}{\kappa(s)}(g'(s) \times g'''(s))$$
$$= \left(\frac{1}{\kappa(s)}\right)'(g'(s) \times g''(s)) + \frac{1}{\kappa(s)}(g'(s) \times g'''(s)), \quad s \in J.$$

Hence,

$$-\tau(s) = \mathbf{b}'(s) \cdot \mathbf{n}(s)$$
$$= \left(\frac{1}{\kappa(s)}\right)'(g'(s) \times g''(s) \cdot \mathbf{n}(s))$$
$$+ \left(\frac{1}{\kappa(s)}\right)(g'(s) \times g'''(s) \cdot \mathbf{n}(s))$$
$$= \left(\frac{1}{\kappa(s)}\right)(g'(s) \times g'''(s) \cdot \mathbf{n}(s))$$
$$= \left(\frac{1}{\kappa(s)}\right)^2 ((g'(s) \times g'''(s)) \cdot g''(s))$$
$$= -\left(\frac{1}{\kappa(s)}\right)^2 ((g'''(s) \times g'(s)) \cdot g''(s))$$
$$= -\left(\frac{1}{\kappa(s)}\right)^2 (g'''(s) \cdot (g'(s) \times g''(s)))$$
$$= -\left(\frac{1}{\kappa(s)}\right)^2 ((g'(s) \times g''(s)) \cdot g'''(s)), \quad s \in J.$$

Therefore,

$$\tau(s) = \left(\frac{1}{\kappa(s)}\right)^2 ((g'(s) \times g''(s)) \cdot g'''(s)), \quad s \in J.$$

Exercise 1.18. Let $(I, f = f(t))$ and $(J, g = g(s))$ be two positively oriented equivalent parametric curves with the parameter change $s : I \to J$, $s' > 0$ on I. Prove that they have the same torsion at the corresponding points t and $s(t)$.

Solution. Let $\{\mathbf{t}, \mathbf{n}, \mathbf{b}\}$ and $\{\mathbf{t}_1, \mathbf{n}_1, \mathbf{b}_1\}$ be the Frenet frames for (I, f) and (J, g), respectively. Then

$$\mathbf{b}_1(s(t)) = \mathbf{b}(t),$$
$$\mathbf{n}_1(s(t)) = \mathbf{n}(t),$$

$$f'(t) = g'(s(t))s'(t), \quad t \in I.$$

From the second equation of the Frenet formulae, we find

$$\mathbf{b}'(t) \cdot \mathbf{n}(t) = -|f'(t)|\tau(t), \quad t \in I.$$

Hence,

$$
\begin{aligned}
\tau(t) &= -\left(\frac{1}{|f'(t)|}\right)(\mathbf{b}'(t) \cdot \mathbf{n}(t)) \\
&= -\frac{1}{|g'(s(t))||s'(t)}((\mathbf{b}_1'(s(t))s'(t)) \cdot \mathbf{n}(t)) \\
&= -\left(\frac{1}{|g'(s(t))|}\right)(\mathbf{b}_1'(s(t)) \cdot \mathbf{n}_1(s(t))) \\
&= -(\mathbf{b}_1'(s(t)) \cdot \mathbf{n}_1(s(t))) \\
&= \tau_1(s(t)), \quad t \in I.
\end{aligned}
$$

This completes the proof.

Now, suppose that (I, f) and (J, g) are equivalent with the parameter change $s : I \to J, s' > 0$ on I. Then

$$
\begin{aligned}
f(t) &= g(s(t)), \\
f'(t) &= g'(s(t))s'(t), \\
f''(t) &= g''(s(t))(s'(t))^2 + g'(s(t))s''(t), \\
f'''(t) &= g'''(s(t))(s'(t))^3 + 2g''(s(t))s'(t)s''(t) \\
&\quad + g''(s(t))s'(t)s''(t) + g'(s(t))s'''(t) \\
&= g'''(s(t))(s'(t))^3 + 3g''(s(t))s'(t)s''(t) + g'(s(t))s'''(t), \quad t \in I.
\end{aligned}
$$

Hence,

$$
\begin{aligned}
(f'(t) &\times f''(t)) \cdot f'''(t) \\
&= (((g'(s(t))s'(t)) \times (g''(s(t))(s'(t))^2 + g'(s(t))s''(t))) \\
&\quad \cdot (g'''(s(t))(s'(t))^3 + 3g''(s(t))s'(t)s''(t) + g'(s(t))s'''(t))) \\
&= (s'(t))^3((g'(s(t)) \times g''(s(t))) \\
&\quad \cdot (g'''(s(t))(s'(t))^3 + 3g''(s(t))s'(t)s''(t) + g'(s(t))s'''(t))) \\
&= (s'(t))^6((g'(s(t)) \times g''(s(t))) \cdot g'''(s(t))) \\
&= (s'(t))^6(\kappa(s(t)))^2\tau(s(t))
\end{aligned}
$$

$$= (s'(t))^6 \frac{|f'(t) \times f''(t)|^2}{|f'(t)|^6} \tau(t)$$

$$= |f'(t) \times f''(t)|^2 \tau(t), \quad t \in I,$$

whereupon

$$\tau(t) = \frac{(f'(t) \times f''(t)) \cdot f'''(t)}{|f'(t) \times f''(t)|^2}, \quad t \in I.$$

Example 1.40. Consider the curve in Example 1.38. Using the computations in Examples 1.38 and 1.39, we get

$$f'''(t) = (a \sinh t, a \cosh t, 0), \quad t \in \mathbb{R},$$

and

$$(f'(t) \times f''(t)) \cdot f'''(t) = -a^3(\sinh t)^2 + a^3(\cosh t)^2$$
$$= a^3((\cosh t)^3 - (\sinh t)^3)$$
$$= a^3, \quad t \in \mathbb{R}.$$

Then

$$\tau(t) = \frac{a^3}{(a^2\sqrt{1 + \cosh(2t)})^2}$$
$$= \frac{1}{a(1 + \cosh(2t))}, \quad t \in \mathbb{R}.$$

Exercise 1.19. Let

$$\mathbf{d} = |f'|(\tau\mathbf{t} + \kappa\mathbf{b}).$$

Prove that the Frenet formulae can be written in the form:

$$\mathbf{t}' = \mathbf{d} \times \mathbf{t},$$
$$\mathbf{n}' = \mathbf{d} \times \mathbf{n},$$
$$\mathbf{b}' = \mathbf{d} \times \mathbf{b}.$$

Solution. We have

$$\mathbf{d} \times \mathbf{t} = |f'|(\tau\mathbf{t} + \kappa\mathbf{b}) \times \mathbf{t}$$
$$= |f'|(\kappa(\mathbf{b} \times \mathbf{t}))$$
$$= |f'|(\kappa\mathbf{n})$$
$$= \mathbf{t}',$$

$$\begin{aligned}
\mathbf{d} \times \mathbf{n} &= (|f'|(\tau \mathbf{t} + \kappa \mathbf{b})) \times \mathbf{n} \\
&= |f'|\tau(\mathbf{t} \times \mathbf{n}) + |f'|\kappa(\mathbf{b} \times \mathbf{n}) \\
&= |f'|\tau \mathbf{b} - |f'|\kappa \mathbf{t} \\
&= \mathbf{n}',
\end{aligned}$$

and

$$\begin{aligned}
\mathbf{d} \times \mathbf{b} &= (|f'|(\tau \mathbf{t} + \kappa \mathbf{b})) \times \mathbf{b} \\
&= (|f'|\tau)(\mathbf{t} \times \mathbf{b}) + (|f'|\kappa)(\mathbf{b} \times \mathbf{b}) \\
&= -(|f'|\tau)\mathbf{n} \\
&= \mathbf{b}'.
\end{aligned}$$

This completes the proof.

Definition 1.24. The vector $\mathbf{d}$ is called Darboux vector.

Exercise 1.20. Prove that the support of a biregular curve lies in a plane if and only if its torsion vanishes identically.

Solution. Let (I, f) be a biregular curve.

1. Suppose that (I, f) lies in a plane Π, i. e., $f(I) \subset \Pi$. Then f' and f'' are parallel to this plane. Thus, Π is the osculating plane and then

$$\mathbf{b}(t) = \text{const}, \quad t \in I.$$

Hence,

$$\begin{aligned}
0 &= \mathbf{b}'(t) \\
&= -|f'(t)| \cdot \tau(t) \cdot \mathbf{n}(t), \quad t \in I.
\end{aligned}$$

Therefore,

$$\tau(t) = 0, \quad t \in I.$$

2. Let

$$\tau(t) = 0, \quad t \in I.$$

Then, by the Frenet formulae, it follows that $\mathbf{b}(t) = \text{const} = \mathbf{b}_0, t \in I$. On the other hand,

$$\mathbf{b}(t) = \mathbf{t}(t) \times \mathbf{n}(t), \quad t \in I.$$

Hence, $f'(t) \perp \mathbf{b}_0, t \in I$, and then

$$0 = f'(t) \cdot \mathbf{b}_0, \quad t \in I.$$

Therefore,

$$\left(f(t) \cdot \mathbf{b}_0 \right)'(t) = 0$$

and

$$\left(f(t) - f(t_0) \right) \cdot \mathbf{b}_0 = 0.$$

Thus, the support of $(I, f = f(t))$ is contained in a plane that is perpendicular to $\mathbf{b}_0$. This completes the proof.

Exercise 1.21. Let $(I, f = f(s))$ be a naturally parameterized curve with constant curvature $\kappa_0 > 0$ and let its torsion be 0. Prove that the support of f lies on a circle of radius $1/\kappa_0$.

Solution. Since the torsion of the curve f is 0, we have that the curve f is a plane curve. Introduce

$$f_1 = f + \frac{1}{\kappa_0} \mathbf{n}.$$

Hence,

$$f_1' = f' + \frac{1}{\kappa_0} \mathbf{n}'$$
$$= \mathbf{t} + \frac{1}{\kappa_0}(-\kappa_0 \mathbf{t})$$
$$= \mathbf{t} - \mathbf{t}$$
$$= 0 \quad \text{on } I.$$

Thus, f_1 is a constant vector c and

$$f - c = -\frac{1}{\kappa_0} \mathbf{t},$$

so that

$$|f - c| = \frac{1}{\kappa_0} |\mathbf{t}|$$
$$= \frac{1}{\kappa_0}.$$

This completes the proof.

Exercise 1.22. Let the support of the naturally parameterized curve $(I, f = f(s))$ lie on a sphere with center $(0, 0, 0)$ and radius $a > 0$. Prove that

$$\kappa \geq \frac{1}{a}.$$

Solution. We have

$$|f|^2 = a^2,$$

whereupon

$$f \cdot f' = 0,$$

or

$$f \cdot \mathbf{t} = 0.$$

We differentiate the latter equation and find

$$\begin{aligned} 0 &= f' \cdot \mathbf{t} + f \cdot \mathbf{t}' \\ &= \mathbf{t} \cdot \mathbf{t} + f \cdot \mathbf{t}' \\ &= 1 + f \cdot (\kappa \mathbf{n}) \end{aligned}$$

whereupon

$$-1 = \kappa(f \cdot \mathbf{n}).$$

Note that

$$|f \cdot \mathbf{n}| \leq |f||\mathbf{n}| = a.$$

Therefore,

$$\kappa = \frac{1}{|f \cdot \mathbf{n}|} \geq \frac{1}{a}.$$

This completes the proof.

1.7 Advanced practical problems

Problem 1.1. Prove that the following curves are regular:
1. for $n = 3$, $[a, b] = [0, 10]$,

$$f(t) = (1 + t + t^2, t^3, t), \quad t \in [0, 10];$$

2. (catenary) $n = 2$, $[a, b] = \mathbb{R}$,

$$f(t) = (t, \cosh t), \quad t \in \mathbb{R};$$

3. for $n = 4$, $[a, b] = [-1, 1]$,

$$f(t) = \left(\frac{1}{2 + t}, t^2, t^3, 1 + t + t^2 \right), \quad t \in [-1, 1];$$

4. for $n = 5$, $[a, b] = [-2, 4]$,

$$f(t) = (e^t, \cos t, \sin t, t^2 + t, 1 - t), \quad t \in [-2, 4];$$

5. for $n = 2$, $[a, b] = [0, 1]$,

$$f(t) = (t^2, 1 + t + t^2 + t^3 + t^4), \quad t \in [0, 1].$$

Problem 1.2. Prove that the following curves are equivalent:
1. (shortened cycloid)

$$f(t) = (at - d \sin t, a - d \cos t), \quad t \in \left[\frac{\pi}{4}, \frac{\pi}{2} \right],$$

and

$$g(s) = (a \arcsin s - ds, a - d\sqrt{1 - s^2}), \quad s \in \left[\frac{\sqrt{2}}{2}, 1 \right].$$

2. (lengthened cycloid)

$$f(t) = (at - d \sin t, a - d \cos t), \quad t \in \left[\frac{\pi}{4}, \frac{\pi}{2} \right],$$

and

$$g(t) = (as^3 + a - d \sin(s^3 + 1), a - d \cos(s^3 + 1)), \quad s \in \left[\sqrt[3]{1 - \frac{\pi}{4}}, \sqrt[3]{1 - \frac{\pi}{2}} \right].$$

3. (epicycloid)

$$f(t) = \left((R + r) \cos \left(\frac{r}{R} t \right) - r \cos \left(\frac{R + r}{r} t \right), \right.$$
$$\left. (R + r) \sin \left(\frac{r}{R} t \right) - r \sin \left(\frac{R + r}{r} t \right) \right), \quad t \in \left[\frac{\pi}{4}, \frac{\pi}{2} \right],$$

and

$$g(s) = \left((R+s)\cos\left(\frac{r}{R}s^2\right) - r\cos\left(\frac{R+r}{r}s^2\right),$$

$$(R+r)\sin\left(\frac{r}{R}s^2\right) - r\sin\left(\frac{R+r}{r}s^2\right) \right), \quad s \in \left[\frac{\sqrt{\pi}}{2}, \sqrt{\frac{\pi}{2}}\right].$$

Hint 1.7. Use the following functions:
1. $t = \phi(s) = \arcsin s, \ s \in [\frac{\sqrt{2}}{2}, 1];$
2. $t = \phi(s) = s^3 + 1, \ s \in [\sqrt[3]{1-\frac{\pi}{4}}, \sqrt[3]{1-\frac{\pi}{2}}];$
3. $t = \phi(s) = s^2, \ s \in [\frac{\sqrt{\pi}}{2}, \sqrt{\frac{\pi}{2}}].$

Problem 1.3. Let $c > 0$ be a constant. Find the arc length functions of the following curves:
1.
$$f(t) = \left(t, c \cosh\left(\frac{t}{c}\right) \right), \quad t \in [a,b];$$

2.
$$f(t) = (c(\cos t + t \sin t), c(\sin t - t \cos t)), \quad t \in [a,b];$$

3.
$$f(t) = \left(c\left(\log\left(\tan\left(\frac{t}{2}\right)\right) + \cos t\right), c\sin t \right), \quad t \in [a,b];$$

Answer 1.8. 1.
$$L_f(t,a) = \frac{e^{t/c} - e^{-t/c} - e^{a/c} + e^{-a/c}}{2}, \quad t \in [a,b];$$

2.
$$L_f(t,a) = \frac{c}{2}(t^2 - a^2), \quad t \in [a,b];$$

3.
$$L_f(t,a) = c(\log(\sin t) - \log(\sin a)), \quad t \in [a,b].$$

Problem 1.4. Find $L_f(b,a)$, where
1.
$$f(t) = (t, \log(\cos t)), \quad t \in \left[0, \frac{\pi}{3}\right];$$

2.
$$f(t) = \left(t, \frac{1}{4}t^2 - \frac{1}{2}\log t \right), \quad t \in [1,4];$$

3.
$$f(t) = \left(t - \frac{1}{2}\sinh(2t), 2\cosh t \right), \quad t \in [0,2];$$

4.
$$f(t) = (8ct^3, 3c(2t^2 - t^4)), \quad t \in [0, \sqrt{2}],$$

where $c > 0$ is a given constant.

Answer 1.9. 1. $\log(2 + \sqrt{3});$

2. $\frac{15}{4} + \frac{1}{2}\log 2$;

3. $\frac{1}{2}(\cosh 4 - 1)$;

4. $24c$.

Problem 1.5. Prove that the curves in Problem 1.3 are not naturally parameterized.

Problem 1.6. Find the equation of the tangent line at the corresponding points of the following curves:

1.
$$f(t) = (t^3 - 2t, t^2 + 1), \quad t \in \mathbb{R}, \quad t = 1.$$

2.
$$f(t) = (a(\cos t)^3, a(\sin t)^3), \quad t \in \mathbb{R};$$

3.
$$f(t) = (a(t - \sin t), a(t - \cos t)), \quad t \in \mathbb{R};$$

4.
$$f(t) = \left(t + \frac{1}{2}t^2 - \frac{1}{4}t^4, \frac{1}{2}t^2 + \frac{1}{3}t^3\right), \quad t \in \mathbb{R}, \quad t = 0;$$

5.
$$f(t) = (a\cos t, b\sin t), \quad t \in \mathbb{R};$$

6.
$$f(t) = \left(\frac{a}{2}\left(t + \frac{1}{t}\right), \frac{b}{2}\left(t - \frac{1}{t}\right)\right), \quad t \neq 0, \quad t \in \mathbb{R};$$

7.
$$f(t) = (e^t \cos t, e^t \sin t, e^t), \quad t \in \mathbb{R}, \quad t = 0;$$

Answer 1.10. 1.
$$2F_1(\lambda) - F_2(\lambda) + 4 = 0, \quad \lambda \in \mathbb{R};$$

2.
$$2F_1(\lambda)\sin t + 2F_2(\lambda)\cos t - a\sin(2t) = 0, \quad \lambda, t \in \mathbb{R};$$

3.
$$F_2(\lambda) = 2a, \quad t = (2k + 1)\pi, \quad k \in \mathbb{Z}, \quad \lambda \in \mathbb{R},$$

and
$$F_1(\lambda) - F_2(\lambda)\tan\left(\frac{t}{2}\right) + a\left(2\tan\left(\frac{t}{2}\right) - t\right) = 0, \quad \lambda \in \mathbb{R},$$

$t \neq (2k + 1)\pi, \quad k \in \mathbb{Z}$;

4.
$$F_2(\lambda) = 0, \quad \lambda \in \mathbb{R};$$

5.
$$b\cos tF_1(\lambda) + a\sin tF_2(\lambda) - ab = 0, \quad t, \lambda \in \mathbb{R};$$

6.
$$\frac{b}{2}\left(1+\frac{1}{t^2}\right)F_1(\lambda) - \frac{a}{2}\left(1-\frac{1}{t^2}\right)F_2(\lambda) - \frac{ab}{t} = 0, \quad t,\lambda \in \mathbb{R};$$

7.
$$F_1(\lambda) + F_3(\lambda) - 2F_2(\lambda) - 2 = 0, \quad \lambda \in \mathbb{R}.$$

Problem 1.7. Find the equation of the normal plane at $t = 1$ for the following curve:

$$f(t) = (t, t^2, t^3), \quad t \in \mathbb{R}.$$

Answer 1.11.

$$x + 2y + 3z - 6 = 0.$$

Problem 1.8. Prove that following curves are biregular:
1. $f(t) = (e^t, e^{2t}, e^{3t}), t \in [2,4]$;
2. $f(t) = (e^{-t}, e^{-2t}, e^{-7t}), t \in [1,3]$;
3. $f(t) = (t + 7, 2t, t^2), t \in [1,8]$.

Problem 1.9. Find the osculating plane at an arbitrary point for the following curve $t_0 \in \mathbb{R}$:

$$f(t) = (t, t^2, t^3), \quad t \in \mathbb{R}.$$

Answer 1.12.

$$3t_0^2 x - 3t_0 y + z - t_0^3 = 0, \quad t_0 \in \mathbb{R}.$$

Problem 1.10. Find the curvature of the following curves:
1.
$$f(t) = (a(t - \sin t), a(1 - \cos t)), \quad t \in \mathbb{R},$$

where $a \in \mathbb{R}$ is a parameter, $a \neq 0$;
2.
$$f(t) = (2\cos t - \cos(2t), 2\sin t - \sin(2t)), \quad t \in \mathbb{R};$$

3.
$$f(t) = (a(\cos t)^3, a(\sin t)^3), \quad t \in \mathbb{R},$$

where $a \in \mathbb{R}$ is a parameter, $a \neq 0$;
4.
$$f(t) = (a(\cos t + t \sin t), a(\sin t - t \cos t)), \quad t \in \mathbb{R},$$

where $a \in \mathbb{R}$ is a real parameter, $a \neq 0$;
5.
$$f(t) = (e^t, e^{-t}, t\sqrt{2}), \quad t \in \mathbb{R};$$

6.
$$f(t) = (2t, \log t, t^2), \quad t \in \mathbb{R}, \quad t > 0;$$

7.
$$f(t) = ((\cos t)^3, (\sin t)^3, \cos(2t)), \quad t \in \mathbb{R}.$$

Answer 1.13. 1.
$$\kappa(t) = \frac{1}{4a|\sin(\frac{t}{2})|}, \quad t \in \mathbb{R};$$

2.
$$\kappa(t) = \frac{3}{16|\sin(\frac{t}{2})|}, \quad t \in \mathbb{R};$$

3.
$$\kappa(t) = \frac{2}{3|a\sin(2t)|}, \quad t \in \mathbb{R};$$

4.
$$\kappa(t) = \frac{1}{|at|}, \quad t \in \mathbb{R};$$

5.
$$\kappa(t) = \frac{\sqrt{2}}{(e^t + e^{-t})^2}, \quad t \in \mathbb{R};$$

6.
$$\kappa(t) = \frac{2t}{(1 + 2t^2)^3}, \quad t \in \mathbb{R}, \quad t > 0;$$

7.
$$\kappa(t) = \frac{3}{25 \sin t \cos t}, \quad t \in \mathbb{R}.$$

Problem 1.11. Find the torsion of the curves 5, 6, and 7 in Problem 1.10.

Answer 1.14. 1.
$$\tau(t) = -\frac{\sqrt{2}}{(e^t + e^{-t})^2}, \quad t \in \mathbb{R};$$

2.
$$\tau(t) = -\frac{2t}{(1 + 2t^2)^2}, \quad t \in \mathbb{R};$$

3.
$$\tau(t) = \frac{4}{25 \sin t \cos t}, \quad t \in \mathbb{R}.$$

Problem 1.12. Prove that for the following curves the curvature and torsion are equal:

1.
$$f(t) = (a \cosh t, a \sinh t, at), \quad t \in \mathbb{R};$$

2.
$$f(t) = (3t - t^3, 3t^2, 3t + t^3), \quad t \in \mathbb{R}.$$

Problem 1.13. Find the values of the parameters a and b so that for the curve

$$f(t) = (a \cosh t, a \sinh t, bt), \quad t \in \mathbb{R},$$

the curvature and torsion are equal.

Answer 1.15. $a = b$.

Problem 1.14. Find the points on the curve

$$f(t) = ((\cos t)^3, (\sin t)^3, \cos(2t)), \quad t \in \mathbb{R},$$

at which the curvature and torsion are equal.

Answer 1.16.

$$t = \frac{\pi}{4} + k\pi, \quad k \in \mathbb{Z}.$$

Definition 1.25. A parameterized curve (I, f) is said to be a general helix if its tangents make a constant angle with a fixed vector in $\mathbb{R}^3$.

Problem 1.15 (Lancret theorem). Prove that a space curve (I, f) with the curvature $\kappa > 0$ is a general helix if and only if the ratio of its torsion and curvature is a constant.

Solution. Without loss of generality, suppose that (I, f) is a naturally parameterized curve.

1. Let (I, f) be a general helix and $\vec{v}$ be a fixed direction that makes a constant angle with its tangents. Then

$$\mathbf{t} \cdot \vec{v} = \cos \alpha_0 = \text{const}.$$

Hence, differentiating we arrive at

$$\mathbf{t}' \cdot \vec{v} = 0,$$

whereupon

$$\kappa(\mathbf{n} \cdot \vec{v}) = 0.$$

Since $\kappa > 0$, we find

$$\mathbf{n} \cdot \vec{v} = 0. \tag{1.9}$$

Thus, $\vec{v} \perp \mathbf{n}$ and

$$\mathbf{b} \cdot \vec{v} = \sin \alpha_0.$$

Now, we differentiate equation (1.9) and obtain

$$\begin{aligned}
0 &= \mathbf{n}' \cdot \vec{v} \\
&= (-\kappa \mathbf{t} + \tau \mathbf{b}) \cdot \vec{v} \\
&= -\kappa(\mathbf{t} \cdot \vec{v}) + \tau(\mathbf{b} \cdot \vec{v}) \\
&= -\kappa \cos \alpha_0 + \tau \sin \alpha_0,
\end{aligned}$$

from where

$$\frac{\tau}{\kappa} = \frac{\cos \alpha_0}{\sin \alpha_0}$$
$$= \cot \alpha_0$$
$$= \text{const.}$$

2. Let

$$\frac{\tau}{\kappa} = c_0,$$

where c_0 is a constant. Then

$$\tau = c_0 \kappa,$$

or

$$c_0 \kappa - \tau = 0.$$

Hence,

$$0 = (c_0 \kappa - \tau)\mathbf{n}$$
$$= c_0(\kappa \mathbf{n}) - \tau \mathbf{n}$$
$$= c_0 \mathbf{t}' + \mathbf{b}',$$

whereupon, after integrating,

$$c_0 \mathbf{t} + \mathbf{b} = \vec{w}, \quad 0 \neq \vec{w} \in \mathbb{R}^3.$$

Let

$$\vec{v} = \frac{1}{|\vec{w}|}\vec{w}.$$

Then

$$\vec{v} = \frac{1}{|c_0 \mathbf{t} + \mathbf{b}|}(c_0 \mathbf{t} + \mathbf{b})$$
$$= \frac{1}{(1 + c_0^2)^{\frac{1}{2}}}(c_0 \mathbf{t} + \mathbf{b}).$$

Hence,

$$\vec{v} \cdot \mathbf{t} = \frac{c_0}{(1 + c_0^2)^{\frac{1}{2}}}.$$

Thus, $\vec{v}$ and $\mathbf{t}$ make a constant angle. Therefore (I, f) is a general helix. This completes the proof.

Definition 1.26. The curves that have the same principal normals are said to be Bertrand curves. Usually, for a Bertrand curve there is only one curve having the same principal normals. The two curves are said to be Bertrand mates, or associated Bertrand curves, or conjugated Bertrand curves. If a Bertrand curve has more than one mate, then it has infinitely many, and the curve and its Bertrand mates are said to be a circular cylindrical helix.

Let f_1 and f_2 be Bertrand mates and f_1 be naturally parameterized with the parameter change s. Then f_2 depends on s and we assume that $f_2(s)$ and $f_1(s)$ have the same principal normals. Both points will be called corresponding points.

Exercise 1.23 (Shell theorem). Suppose that f_1 and f_2 are two associated Bertrand curves and f_1 is naturally parameterized with the parameter change s. Prove that the angle of the tangents at the corresponding points is a constant.

Solution. Let $I \subseteq \mathbb{R}$ be the range of s. Let also $\{\mathbf{t}_1, \mathbf{n}_1, \mathbf{b}_1\}$ and $\{\mathbf{t}_2, \mathbf{n}_2, \mathbf{b}_2\}$ be the Frenet frames of f_1 and f_2, respectively. Then

$$f_2(s) = f_1(s) + a(s)\mathbf{n}_1(s), \quad s \in I, \tag{1.10}$$

for some $a \in \mathcal{C}^1(I)$. For the principal normals $\mathbf{n}_1$ and $\mathbf{n}_2$, we have the relations

$$\mathbf{n}_2(s) = \pm\mathbf{n}_1(s), \quad s \in I. \tag{1.11}$$

We differentiate equation (1.10) with respect to s and find

$$\frac{df_2}{ds}(s) = \frac{df_1}{ds}(s)\frac{da}{ds}(s)\mathbf{n}_1(s) + a(s)\frac{d\mathbf{n}_1}{ds}(s)$$

$$= \mathbf{t}_1(s) + \frac{da}{ds}(s)\mathbf{n}_1(s) + a(s)\left(-\kappa_1(s)\mathbf{t}_1(s) + \tau_1(s)\mathbf{b}_1(s)\right)$$

$$= \left(1 - a(s)\kappa_1(s)\right)\mathbf{t}_1(s) + \frac{da}{ds}(s)\mathbf{n}_1(s) + a(s)\tau_1(s)\mathbf{b}_1(s), \quad s \in I.$$

Since

$$\frac{df_2}{ds}(s), \quad s \in I,$$

is a tangent vector to f_2 and $s \in I$, it is perpendicular to $\mathbf{n}_1(s)$ and $\mathbf{n}_2(s)$, $s \in I$. Hence using the latter equation, we find

$$0 = \frac{df_2}{ds}(s) \cdot \mathbf{n}_1(s)$$

$$= \frac{da}{ds}(s), \quad s \in I.$$

Therefore, a is a constant on I and (1.10) can be written in the form

$$f_2(s) = f_1(s) + a\mathbf{n}_1(s), \quad s \in I, \tag{1.12}$$

and

$$\frac{df_2}{ds}(s) = (1 - a\kappa_1(s))\mathbf{t}_1(s) + a\tau_1(s)\mathbf{b}_1(s), \quad s \in I. \tag{1.13}$$

Let now s_2 be the arc length parameter of f_2. Then

$$
\begin{aligned}
\mathbf{t}_2(s) &= \frac{df_2}{ds_2}(s) \\
&= \frac{df_2}{ds}(s)\frac{ds}{ds_2}(s) \tag{1.14} \\
&= (1 - a\kappa_1(s))\mathbf{t}_1(s)\frac{ds}{ds_2}(s) + a\tau_1(s)\mathbf{b}_1(s)\frac{ds}{ds_2}(s), \quad s \in I.
\end{aligned}
$$

Let

$$w(s) = \angle(\mathbf{t}_1(s), \mathbf{t}_2(s)), \quad s \in I.$$

Then

$$\mathbf{t}_1(s) \cdot \mathbf{t}_2(s) = \cos w(s), \quad s \in I.$$

Hence,

$$\mathbf{t}_2(s) = \cos w(s)\mathbf{t}_1(s) + \sin w(s)\mathbf{b}_1(s), \quad s \in I, \tag{1.15}$$

and

$$\mathbf{b}_2(s) = \varepsilon\left(-\sin w(s)\mathbf{t}_1(s) + \cos w(s)\mathbf{b}_1(s)\right), \quad s \in I,$$

where $\varepsilon = \pm 1$. We differentiate with respect to s equation (1.15) and find

$$
\begin{aligned}
\frac{d\mathbf{t}_2}{ds}(s) &= -\sin w(s)\frac{dw}{ds}(s)\mathbf{t}_1(s) + \cos w(s)\frac{d\mathbf{t}_1}{ds} \\
&\quad + \cos w(s)\frac{dw}{ds}(s)\mathbf{b}_1(s) + \sin w(s)\frac{d\mathbf{b}_1}{ds}(s) \\
&= -\sin w(s)\frac{dw}{ds}(s)\mathbf{t}_1(s) + \cos w(s)\kappa_1(s)\mathbf{n}_1(s) \\
&\quad + \cos w(s)\frac{dw}{ds}(s)\mathbf{b}_1(s) - \sin w(s)\tau_1(s)\mathbf{n}_1(s), \quad s \in I.
\end{aligned}
$$

Since

$$\frac{dt_2}{ds}(s) = \frac{dt_2}{ds_2}(s_2)\frac{ds_2}{ds}(s),$$

$$= \kappa_2 n_2 \frac{ds_2}{ds}(s), \quad s \in I,$$

we have that

$$\frac{dt_2}{ds}(s), \quad s \in I,$$

is collinear with $n_2(s)$, $s \in I$. Therefore

$$0 = \frac{dt_2}{ds}(s) \cdot t_1(s),$$

$$0 = \frac{dt_2}{ds}(s) \cdot b_1(s), \quad s \in I,$$

and then

$$\frac{dw}{ds}(s) = 0, \quad s \in I.$$

Thus, w is a constant on I. This completes the solution.

Problem 1.16 (Bertrand theorem). Let f_1 be a naturally parameterized curve on I with parameter change s. Prove that f_1 is a Bertrand curve if and only if its torsion and curvature satisfy

$$a\kappa_1(s) + b\tau_1(s) = 1, \quad s \in I, \tag{1.16}$$

where a and b are constants.

Solution. We will use the notations used in the solution of Exercise 1.23.
1. Let f_1 and f_2 be Bertrand mates. Thus, from (1.14) and (1.15), we find

$$\cos w = (1 - a\kappa_1(s))\frac{ds}{ds_2}(s),$$

$$\sin w = a\tau_1(s)\frac{ds}{ds_2}(s), \quad s \in I,$$

whereupon

$$\cot w = \frac{1 - a\kappa_1(s)}{a\tau_1(s)}, \quad s \in I,$$

or

$$(a\tau_1(s))\cot w = 1 - a\kappa_1(s), \quad s \in I.$$

Let

$$b = a \cot w.$$

Hence,

$$a\kappa_1(s) + b\tau_1(s) = 1, \quad s \in I,$$

i.e., we get (1.16).

2. Suppose that (1.16) holds. Then

$$\frac{df_2}{ds}(s) = (1 - a\kappa_1(s))\mathbf{t}_1(s) + a\tau_1(s)\mathbf{b}_1(s), \quad s \in I.$$

Therefore,

$$\frac{df_2}{ds}(s) \cdot \frac{df_2}{ds}(s) = (\tau_1(s))^2(a^2 + b^2), \quad s \in I.$$

Note that

$$\frac{df_2}{ds}(s) \cdot \frac{df_2}{ds}(s) = \left(\frac{df_2}{ds_2}(s)\frac{ds_2}{ds}(s)\right) \cdot \left(\frac{df_2}{ds}(s)\frac{ds_2}{ds}(s)\right)$$

$$= \left(\frac{ds_2}{ds}(s)\right)^2$$

$$= \frac{1}{(\frac{ds}{ds_2}(s))^2}, \quad s \in I.$$

Consequently,

$$1 = (\tau_1(s))^2\left(\frac{ds}{ds_2}(s)\right)^2(a^2 + b^2), \quad s \in I.$$

From here, we conclude that

$$\tau_1(s)\frac{ds}{ds_2}(s) = \text{const}, \quad s \in I.$$

As above,

$$(1 - a\kappa_1(s))\frac{ds}{ds_2}(s) = \text{const}, \quad s \in I.$$

Now, we differentiate equation (1.14) with respect to the parameter change s and find

$$\kappa_2(s)\mathbf{n}_2(s) = \frac{d\mathbf{t}_2}{ds}(s)$$

$$= (1 - a\kappa_1(s))\frac{ds}{ds_2}(s)\frac{d\mathbf{t}_1}{ds}(s) + a\tau_1(s)\frac{ds}{ds_2}(s)\frac{d\mathbf{b}_1}{ds}(s)$$

$$= (1 - a\kappa_1(s))\frac{ds}{ds_2}(s)\kappa_1(s)\mathbf{n}_1(s) - a\tau_1(s)\frac{ds}{ds_2}(s)\tau_1(s)\mathbf{n}_1(s)$$

$$= ((1 - a\kappa_1(s))\kappa_1(s) - a(\tau_1(s))^2)\frac{ds}{ds_2}(s)\mathbf{n}_1(s), \quad s \in I.$$

Thus, $\mathbf{n}_1$ and $\mathbf{n}_2$ are collinear on I. Hence, we conclude that f_1 and f_2 are Bertrand mates. This completes the proof.

Suppose that $I \subseteq \mathbb{R}$.

Definition 1.27. A rigid motion of $\mathbb{R}^3$ is a map $D : \mathbb{R}^3 \to \mathbb{R}^3$,

$$Dx = Ax + b,$$

where $A \in \mathcal{M}_{3\times3}$, $A^T A = I$, $\det A = 1$, and $b \in \mathbb{R}^3$. The map

$$x \to Ax, \quad x \in \mathbb{R}^3,$$

is said to be the homogeneous part of the motion.

Problem 1.17. Let $(I, f = f(t))$ be a biregular curve and $\{\mathbf{t}(t), \mathbf{n}(t), \mathbf{b}(t)\}$ be its Frenet frame at $t \in I$. Let also $D : \mathbb{R}^3 \to \mathbb{R}^3$ be a rigid motion with homogeneous part A. Prove that

$$\{A\mathbf{t}(t), A\mathbf{n}(t), A\mathbf{b}(t)\}$$

is the Frenet frame of $(I, f_1 = Df)$ at t, and f and f_1 have the same curvature and torsion.

Solution. We have

$$f_1(t) = Af(t) + b.$$

Then

$$f_1'(t) = Af'(t),$$
$$f_1''(t) = Af''(t),$$
$$f_1'''(t) = Af'''(t).$$

Therefore,

$$\begin{aligned}
\mathbf{t}_1(t) &= \frac{f_1'(t)}{|f_1'(t)|} \\
&= \frac{Af'(t)}{|Af'(t)|} \\
&= \frac{Af'(t)}{|f'(t)|}
\end{aligned}$$

$$= A\frac{f'(t)}{|f'(t)|}$$

$$= A\mathbf{t}(t), \quad t \in I,$$

$$\kappa_1(t) = \frac{|f_1'(t) \times f_1''(t)|}{|f_1'(t)|^3}$$

$$= \frac{|Af'(t) \times Af''(t)|}{|Af'(t)|^3}$$

$$= \frac{|f'(t) \times f''(t)|}{|f'(t)|^3}$$

$$= \kappa(t), \quad t \in I,$$

and

$$\mathbf{k}_1(t) = \frac{f_1''(t)}{|f_1'(t)|^3} - \frac{(f_1'(t) \cdot f_1''(t))f_1'(t)}{|f_1'|^4}$$

$$= \frac{Af''(t)}{|Af'(t)|^2} - \frac{(Af'(t) \cdot Af''(t))\,Af'(t)}{|Af'(t)|^4}$$

$$= A\frac{f''(t)}{|f'(t)|^2} - A\frac{(f'(t) \cdot f''(t))f'(t)}{|f'(t)|^4}$$

$$= A\left(\frac{f''(t)}{|f'(t)|^2} - \frac{(f'(t) \cdot f''(t))f'(t)}{|f'(t)|^4}\right)$$

$$= A\mathbf{k}(t), \quad t \in I,$$

as well as

$$\mathbf{n}_1(t) = \frac{\mathbf{k}_1(t)}{\kappa_1(t)}$$

$$= A\frac{\mathbf{k}(t)}{\kappa(t)}$$

$$= A\mathbf{n}(t), \quad t \in I,$$

$$\mathbf{b}_1(t) = \frac{f_1'(t) \times f_1''(t)}{|f_1'(t) \times f_1''(t)|}$$

$$= \frac{(Af'(t)) \times (Af''(t))}{|(Af'(t)) \times (Af''(t))|}$$

$$= \frac{A(f'(t) \times f''(t))}{|f'(t) \times f''(t)|}$$

$$= A\frac{f'(t) \times f''(t)}{|f'(t) \times f''(t)|}$$

$$= A\mathbf{b}(t), \quad t \in I,$$

and

$$\tau_1(t) = \frac{(f_1'(t) \times f_1''(t)) \cdot f_1'''(t)}{|f_1'(t) \times f_1''(t)|^2}$$

$$= \frac{((Af'(t)) \times (Af''(t))) \cdot (A \times f'''(t))}{|(Af'(t)) \times (Af''(t))|^2}$$

$$= \frac{(f'(t) \times f''(t)) \cdot f'''(t)}{|f'(t) \times f''(t)|^2}$$

$$= \tau(t), \quad t \in I.$$

This completes the proof.

Problem 1.18. Let $m, n \in C^1(I)$, $m > 0$ on I. Prove that there is a unique parameterized curve $(I, f = f(s))$ for which

$$\kappa(s) = m(s) \quad \text{and} \quad \tau(s) = n(s), \quad s \in I.$$

Solution. Let $\{e_1, e_2, e_3\}$ be a frame in $\mathbb{R}^3$ at $t_0 \in \mathbb{R}$. Consider the system

$$\mathbf{x}_1' = m(s)\mathbf{x}_2,$$
$$\mathbf{x}_2' = -m(s)\mathbf{x}_1 + n(s)\mathbf{x}_3,$$
$$\mathbf{x}_3' = -n(s)\mathbf{x}_2, \quad s \in I.$$

Define the matrix

$$A(s) = \begin{pmatrix} 0 & m(s) & 0 \\ -m(s) & 0 & n(s) \\ 0 & -n(s) & 0 \end{pmatrix}, \quad s \in I.$$

Then we get the Cauchy problem

$$\mathbf{x}' = A(s)\mathbf{x}, \quad s \in I,$$
$$\mathbf{x}(s_0) = (e_1, e_2, e_3).$$

The last Cauchy problem has a unique solution $\mathbf{x}$. For it, we have

$$(\mathbf{x}^T\mathbf{x})' = (\mathbf{x}^T)'\mathbf{x} + \mathbf{x}^T\mathbf{x}'$$
$$= \mathbf{x}^T A^T \mathbf{x} + \mathbf{x}^T A\mathbf{x}$$
$$= \mathbf{x}^T(A^T + A)\mathbf{x}$$
$$= 0.$$

Thus,

$$(\mathbf{x}^T\mathbf{x}) = \text{const}.$$

Since

$$\mathbf{x}^T(s_0)\mathbf{x}(s_0) = I,$$

we get

$$\mathbf{x}^T\mathbf{x} = I.$$

Define

$$f(s) = m_0 + \int_{s_0}^{s} \mathbf{x}_1(t)dt, \quad s \in I.$$

We have

$$f'(s) = \mathbf{x}_1(s),$$
$$|f'(s)| = |\mathbf{x}_1(s)| = 1,$$
$$f''(s) = \mathbf{x}_1'(s) = m(s)\mathbf{x}_2(s), \quad s \in I.$$

Hence,

$$f'(s) \times f''(s) = m(s)(\mathbf{x}_1(s) \times \mathbf{x}_2(s)) \neq 0, \quad s \in I.$$

Therefore,

$$
\begin{aligned}
f'''(s) &= m'(s)\mathbf{x}_2(s) + m(s)\mathbf{x}_2'(s) \\
&= m'(s)\mathbf{x}_2(s) + m(s)(-m(s)\mathbf{x}_1(s) + n(s)\mathbf{x}_3(s)) \\
&= -(m(s))^2\mathbf{x}_1(s) + m'(s)\mathbf{x}_2(s) + (m(s)n(s))\mathbf{x}_3(s), \quad s \in I.
\end{aligned}
$$

From here,

$$
\begin{aligned}
((f'(s) \times f''(s)) \cdot f'''(s)) &= ((m(s)(\mathbf{x}_1(s) \times \mathbf{x}_2(s))) \\
&\quad \cdot (-(m(s))^2\mathbf{x}_1(s) + m'(s)\mathbf{x}_2(s) + (m(s)n(s))\mathbf{x}_3(s))) \\
&= (m(s))^2 n(s)((\mathbf{x}_1(s) \times \mathbf{x}_2(s)) \cdot \mathbf{x}_3(s)) \\
&= (m(s))^2 n(s), \quad s \in I.
\end{aligned}
$$

Therefore

$$
\begin{aligned}
\kappa(s) &= \frac{|f'(s) \times f''(s)|}{|f'(s)|^3} \\
&= m(s), \quad s \in I,
\end{aligned}
$$

and

$$\tau(s) = \frac{(f'(s) \times f''(s)) \cdot f'''(s)}{|f'(s) \times f''(s)|^2}$$
$$= \frac{(m(s))^2 n(s)}{(m(s))^2}$$
$$= n(s), \quad s \in I.$$

This completes the solution.

Problem 1.19. Let (I, f) and (I, g) be two biregular parameterized curves and let

$$\kappa(t) = \kappa_1(t), \quad \tau(t) = \tau_1(t), \quad |f'(t)| = |g'(t)|, \quad t \in I.$$

Prove that there is a rigid motion $D : \mathbb{R}^3 \to \mathbb{R}^3$ such that

$$g = Df.$$

Solution. Let $t_0 \in I$ and $\{\mathbf{t}, \mathbf{n}, \mathbf{b}\}$, $\{\mathbf{t}_1, \mathbf{n}_1, \mathbf{b}_1\}$ be the Frenet frames of f and g, respectively at $t \in I$. Let also $D : \mathbb{R}^3 \to \mathbb{R}^3$ be the rigid motion such that

$$D\{\mathbf{t}, \mathbf{n}, \mathbf{b}\} = \{\mathbf{t}_1, \mathbf{n}_1, \mathbf{b}_1\}.$$

Let $f_2 = Df$ and $\{\mathbf{t}_2, \mathbf{n}_2, \mathbf{b}_2\}$ be the Frenet frame of f_2. We have

$$\kappa_2(t) = \kappa(t) = \kappa_1(t),$$
$$\tau_2(t) = \tau(t) = \tau_1(t),$$
$$|f_2'(t)| = |g'(t)|, \quad t \in I.$$

Thus, $\{\mathbf{t}_2, \mathbf{n}_2, \mathbf{b}_2\}$ and $\{\mathbf{t}_1, \mathbf{n}_1, \mathbf{b}_1\}$ satisfy the system

$$\mathbf{t}' = |f'|\kappa\mathbf{n},$$
$$\mathbf{n}' = -|f'|\kappa\mathbf{t} + |f'|\tau\mathbf{b},$$
$$\mathbf{b}' = -|f'|\tau\mathbf{n}.$$

Since for $f = f_0$, these solutions coincide, i. e.,

$$\mathbf{t}_2 = \mathbf{t}_1,$$
$$\mathbf{n}_2 = \mathbf{n}_1,$$
$$\mathbf{b}_2 = \mathbf{b}_1 \quad \text{on } I.$$

Hence,

$$\frac{f_2'(t)}{|f_2'(t)|} = \frac{g'(t)}{|g'(t)|}, \quad t \in I,$$

whereupon

$$f_2'(t) = g'(t), \quad t \in I,$$
$$f_2(t) - g(t) = \text{const}, \quad t \in I,$$

and

$$f_2(t) - g(t) = f_2(t_0) - g(t_0) = 0, \quad t \in I.$$

This completes the proof.

2 Plane curves

In this chapter, we are interested in a particular class of curves, the plane curves. We first introduce envelopes of families of such curves in $\mathbb{R}^2$ which depend on a single parameter. Then, the evolutes of plane curves are defined as envelopes of their normals. A key notion in $\mathbb{R}^2$ is the complex structure J for studying the curvature of plane curves. The notions of rotation angle and signed curvature are also explored.

2.1 Envelopes of plane curves

Suppose that $I, J, A \subseteq \mathbb{R}$ and

$$f = f(t, \lambda), \quad t \in I, \quad \lambda \in A. \tag{2.1}$$

Definition 2.1. The envelope of the family (2.1) is a parameterized curve tangent to a member of the family at each point.

Exercise 2.1. Prove that the envelope of the family (2.1) is subject to the equations

$$f = f(t, \lambda)$$

and

$$f_t \times f_\lambda = 0.$$

Solution. Let (J, g) be the envelope of the family (2.1) and $P \in g$. Then P is a tangency point between g and a member of the family (2.1). Thus, its equation depends on λ, i.e.,

$$g = g(\lambda), \quad \lambda \in A.$$

Since P lies on a curve of the family (2.1),

$$g = f(t(\lambda), \lambda).$$

The tangency condition between g and $f(t, \lambda)$ is as follows:

$$g_\lambda \parallel f_t,$$

whereupon

$$g_\lambda \times f_t = 0.$$

Hence, using that

https://doi.org/10.1515/9783111501857-002

$$g_\lambda = f_t t' + f_\lambda,$$

we get

$$0 = (f_t t' + f_\lambda) \times f_t$$
$$= f_\lambda \times f_t.$$

Let $f = (f_1, f_2)$. In fact, we have

$$f_t \times f_\lambda = (0, 0, f_{1\lambda} f_{2t} - f_{1t} f_{2\lambda}).$$

Hence,

$$0 = f_\lambda \times f_t$$

is equivalent to the condition

$$f_{1\lambda} f_{2t} - f_{1t} f_{2\lambda} = 0,$$

which is the equation of the classical envelope of the family (2.1). This completes the proof.

Example 2.1. Let $f : \mathbb{R}^2 \to \mathbb{R}^2$ be given by

$$f(t, \lambda) = (t^2 + \lambda t + \lambda^2, t^2 - \lambda t + \lambda^2), \quad (t, \lambda) \in \mathbb{R}^2.$$

We have

$$f_1(t, \lambda) = t^2 + \lambda t + \lambda^2,$$
$$f_2(t, \lambda) = t^2 - \lambda t + \lambda^2, \quad (t, \lambda) \in \mathbb{R}^2.$$

Then

$$f_{1\lambda}(t, \lambda) = t + 2\lambda,$$
$$f_{2\lambda}(t, \lambda) = -t + 2\lambda,$$
$$f_{1t}(t, \lambda) = 2t + \lambda,$$
$$f_{2t}(t, \lambda) = 2t - \lambda, \quad (t, \lambda) \in \mathbb{R}^2.$$

Hence,

$$f_t(t, \lambda) = (f_{1t}(t, \lambda), f_{2t}(t, \lambda))$$
$$= (2t + \lambda, 2t - \lambda),$$
$$f_\lambda(t, \lambda) = (f_{1\lambda}(t, \lambda), f_{2\lambda}(t, \lambda))$$
$$= (t + 2\lambda, -t + 2\lambda), \quad (t, \lambda) \in \mathbb{R}^2,$$

and the equation of the envelope for the considered family is (see Fig. 2.1)

$$
\begin{aligned}
0 &= (t + 2\lambda)(2t - \lambda) - (2t + \lambda)(-t + 2\lambda) \\
&= 2t^2 - t\lambda + 4\lambda t - 2\lambda^2 - (-2t^2 + 4\lambda t - \lambda t + 2\lambda^2) \\
&= 2t^2 + 3\lambda t - 2\lambda^2 + 2t^2 - 3\lambda t - 2\lambda^2 \\
&= 4t^2 - 4\lambda^2 \\
&= 4(t - \lambda)(t + \lambda), \quad (t, \lambda) \in \mathbb{R}^2,
\end{aligned}
$$

or

$$
t = \pm\lambda, \quad (t, \lambda) \in \mathbb{R}^2.
$$

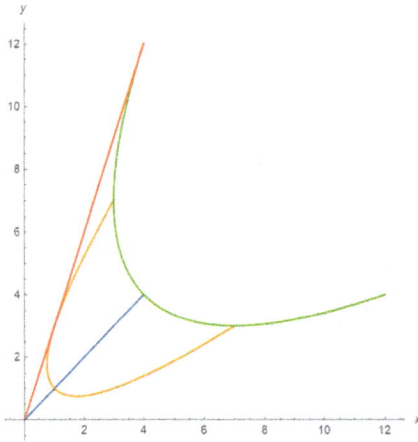

Figure 2.1: The members of the family $f(t, \lambda) = (t^2 + \lambda t + \lambda^2, t^2 - \lambda t + \lambda^2)$, $t \in [-2, 2]$, and its envelope: (blue) $\lambda = 0$; (orange) $\lambda = 1$; (green) $\lambda = 2$; and (red) $\lambda = t$.

Example 2.2. Let $f : \mathbb{R}^2 \to \mathbb{R}^2$ be given by

$$
f(t, \lambda) = (t + \lambda, t - \lambda), \quad (t, \lambda) \in \mathbb{R}^2;
$$

see Fig 2.2. We have

$$
\begin{aligned}
f_1(t, \lambda) &= t + \lambda, \\
f_2(t, \lambda) &= t - \lambda, \quad (t, \lambda) \in \mathbb{R}^2.
\end{aligned}
$$

Hence,

$$
\begin{aligned}
f_{1t}(t, \lambda) &= 1, \\
f_{2t}(t, \lambda) &= 1,
\end{aligned}
$$

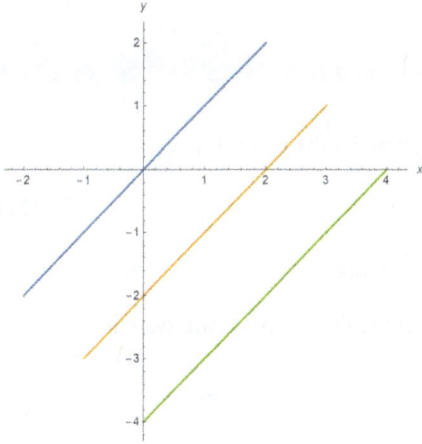

Figure 2.2: The members of the family $f(t,\lambda) = (t + \lambda t, t - \lambda), t \in [-2,2]$: (blue) $\lambda = 0$; (orange) $\lambda = 1$; (green) $\lambda = 2$.

$$f_{1\lambda}(t,\lambda) = 1,$$
$$f_{2\lambda}(t,\lambda) = -1, \quad (t,\lambda) \in \mathbb{R}^2,$$

and

$$f_t(t,\lambda) = (f_{1t}(t,\lambda), f_{2t}(t,\lambda))$$
$$= (1,1),$$
$$f_\lambda(t,\lambda) = (f_{1\lambda}(t,\lambda), f_{2\lambda}(t,\lambda))$$
$$= (1,-1), \quad (t,\lambda) \in \mathbb{R}^2,$$

so the equation of the envelope is

$$0 = f_{1t}(t,\lambda)f_{2\lambda}(t,\lambda) - f_{1\lambda}(t,\lambda)f_{2t}(t,\lambda)$$
$$= 1 \cdot 1 - 1 \cdot (-1)$$
$$= 1 + 1$$
$$= 2,$$

which is impossible. Thus, the considered family of curves has no envelope.

Exercise 2.2. Let $f : \mathbb{R}^2 \to \mathbb{R}^2$ be given by

$$f(t,\lambda) = (\lambda + a\cos t, \lambda + a\sin t), \quad (t,\lambda) \in \mathbb{R}^2,$$

where $a > 0$ is a given constant. Find the equation of the envelope of the considered family.

Answer 2.1.

$$\left(\lambda \pm \frac{a}{\sqrt{2}}, \lambda \mp \frac{a}{\sqrt{2}}\right), \quad \lambda \in \mathbb{R}.$$

Now, suppose that the family of curves is given by the equation

$$F(x, y, \lambda) = 0, \tag{2.2}$$

where F is a C^1-function with respect to its arguments.

Exercise 2.3. Prove that the envelope for the family (2.2) satisfies the system

$$F(x, y, \lambda) = 0,$$
$$F_\lambda(x, y, \lambda) = 0. \tag{2.3}$$

Solution. Locally, a curve of the family around some point can be represented in the form

$$x = x(t, \lambda),$$
$$y = y(t, \lambda),$$

and equation (2.2) can be written as follows:

$$F(x(t, \lambda), y(t, \lambda), \lambda) = 0. \tag{2.4}$$

Let

$$f(t, \lambda) = (x(t, \lambda), y(t, \lambda)).$$

We have

$$f_t(t, \lambda) = (x_t(t, \lambda), y_t(t, \lambda)),$$
$$f_\lambda(t, \lambda) = (x_\lambda(t, \lambda), y_\lambda(t, \lambda)).$$

From the equation

$$f_t \times f_\lambda = 0,$$

we get

$$x_t(t, \lambda) y_\lambda(t, \lambda) - x_\lambda(t, \lambda) y_t(t, \lambda) = 0,$$

and then there is a constant $K \in \mathbb{R}$ such that

$$x_\lambda(t, \lambda) = K x_t(t, \lambda),$$
$$y_\lambda(t, \lambda) = K y_t(t, \lambda).$$

Now, we differentiate equation (2.4) with respect to t and λ and find

$$F_x x_t + F_y y_t = 0,$$
$$F_x x_\lambda + F_y y_\lambda + F_\lambda = 0,$$

whereupon

$$0 = F_x K x_t + F_y K y_t + F_\lambda$$
$$= K(F_x x_t + F_y y_t) + F_\lambda$$
$$= F_\lambda.$$

This completes the solution.

Example 2.3. Consider the following family of curves:

$$F(x,y,\lambda) = (x - \lambda)^2 + (y - \lambda)^2 - a^2 = 0, \quad (x,y) \in \mathbb{R}^2, \quad \lambda \in \mathbb{R},$$

where $a > 0$ is a given parameter. See Fig. 2.3 for the value of $a = 1$. We have

$$F_\lambda(x,y,\lambda) = -2(x - \lambda) - 2(y - \lambda), \quad (x,y) \in \mathbb{R}^2, \quad \lambda \in \mathbb{R}.$$

Hence, we get the system

$$(x - \lambda)^2 + (y - \lambda)^2 - a^2 = 0,$$
$$-2(x - \lambda) - 2(y - \lambda) = 0, \quad (x,y) \in \mathbb{R}^2, \quad \lambda \in \mathbb{R},$$

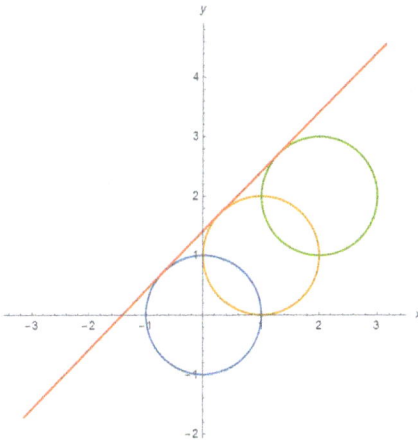

Figure 2.3: The members of the family $F(x,y,\lambda) = (x - \lambda)^2 + (y - \lambda)^2 - a^2 = 0$, where $a = 1$ and $t \in [-\pi, \pi]$: (blue) $\lambda = 0$; (orange) $\lambda = 1$; (green) $\lambda = 2$; and (red) $\lambda = (x + y)/2$.

or

$$(x - \lambda)^2 + (y - \lambda)^2 - a^2 = 0,$$
$$x + y - 2\lambda = 0, \quad (x,y) \in \mathbb{R}^2, \quad \lambda \in \mathbb{R},$$

or

$$\lambda = \frac{x + y}{2},$$
$$\left(x - \frac{x + y}{2}\right)^2 + \left(y - \frac{x + y}{2}\right)^2 = a^2, \quad (x,y) \in \mathbb{R}^2, \quad \lambda \in \mathbb{R},$$

whereupon

$$\frac{(x - y)^2}{4} + \frac{(x - y)^2}{4} = a^2, \quad (x,y) \in \mathbb{R}^2$$

or

$$(x - y)^2 = 2a^2, \quad (x,y) \in \mathbb{R}^2.$$

Thus, the envelopes of the considered family of curves are

$$x - y = \pm\sqrt{2}a, \quad (x,y) \in \mathbb{R}^2.$$

Example 2.4. Consider the family of curves

$$F(x,y,\lambda) = (x + y - \lambda)^2 - 2\lambda^2 = 0, \quad (x,y) \in \mathbb{R}^2, \quad \lambda \in \mathbb{R}.$$

We have

$$F_\lambda(x,y,\lambda) = 2(x + y + \lambda) - 4\lambda$$
$$= 2(x + y - \lambda), \quad (x,y) \in \mathbb{R}^2, \quad \lambda \in \mathbb{R}.$$

Hence, we get the system

$$(x + y + \lambda)^2 - 2\lambda^2 = 0,$$
$$2(x + y - \lambda) = 0, \quad (x,y) \in \mathbb{R}^2, \quad \lambda \in \mathbb{R},$$

whereupon

$$\lambda = x + y,$$
$$4(x + y)^2 - 2(x + y)^2 = 0, \quad (x,y) \in \mathbb{R}^2, \quad \lambda \in \mathbb{R}.$$

Thus, the envelope of the considered family is

$$y = -x, \quad (x,y) \in \mathbb{R}^2.$$

Exercise 2.4. Find the envelopes of the following families of curves:

1.
$$F(x,y,\lambda) = (x - \lambda)^2 + (y - \lambda)^2 - \lambda^2 = 0, \quad (x,y) \in \mathbb{R}^2, \quad \lambda \in \mathbb{R};$$

2.
$$F(x,y,\lambda) = x \cos \lambda + y \sin \lambda - p = 0, \quad (x,y) \in \mathbb{R}^2, \quad \lambda \in \mathbb{R},$$

where $p > 0$ is a given constant;

3.
$$F(x,y,\lambda) = y - (x - \lambda)^2 = 0, \quad (x,y) \in \mathbb{R}^2, \quad \lambda \in \mathbb{R};$$

4.
$$F(x,y,\lambda) = y^2 - (x - \lambda)^2 = 0, \quad (x,y) \in \mathbb{R}^2, \quad \lambda \in \mathbb{R}.$$

Answer 2.2. 1.
$$y = 0, \quad (x,y) \in \mathbb{R}^2;$$

2.
$$x^2 + y^2 = p^2, \quad (x,y) \in \mathbb{R}^2;$$

3.
$$y = 0, \quad (x,y) \in \mathbb{R}^2;$$

4.
$$y = 0, \quad (x,y) \in \mathbb{R}^2.$$

2.2 The evolute

Suppose that $I \subseteq \mathbb{R}$.

Definition 2.2. Let $(I, f = f(t))$ be a parameterized curve. The envelope of the family of the normals to f is said to be evolute of f.

Let

$$f(t) = (f_1(t), f_2(t)), \quad t \in I.$$

Suppose that $f \in C^2(I)$. The equation of the normals to f is as follows:

$$(X - f_1(t))f_1'(t) + (Y - f_2(t))f_2'(t) = 0, \quad t \in I,$$

or

$$f_1'(t)X + f_2'(t)Y = f_1(t)f_1'(t) + f_2(t)f_2'(t), \quad t \in I.$$

We differentiate it with respect to t and find

$$f_1''(t)X + f_2''(t)Y = f_1(t)f_1''(t) + f_2(t)f_2''(t) + \left(f_1'(t)\right)^2 + \left(f_2'(t)\right)^2.$$

Therefore the evolute of f satisfies the system

$$f_1'(t)X + f_2'(t)Y = f_1(t)f_1'(t) + f_2(t)f_2'(t),$$
$$f_1''(t)X + f_2''(t)Y = f_1(t)f_1''(t) + f_2(t)f_2''(t) + \left(f_1'(t)\right)^2 + \left(f_2'(t)\right)^2, \quad t \in I.$$

Example 2.5. We will find the evolute of the ellipse

$$f(t) = (a\cos t, b\sin t), \quad t \in [0, 2\pi].$$

Here

$$f_1(t) = a\cos t,$$
$$f_2(t) = a\sin t, \quad t \in [0, 2\pi].$$

Then

$$f_1'(t) = -a\sin t,$$
$$f_2'(t) = a\cos t,$$
$$f_1''(t) = -a\cos t,$$
$$f_2''(t) = -b\sin t, \quad t \in [0, 2\pi].$$

Hence, the equations of the evolute of the ellipse are as follows:

$$-a\sin tX + b\cos tY = a\cos t(-a\sin t) + b\sin t(b\cos t),$$
$$-a\cos tX - b\sin tY = a\cos t(-a\cos t) + b\sin t(-b\sin t)$$
$$+ (a\sin t)^2 + (b\cos t)^2, \quad t \in [0, 2\pi],$$

or

$$-a\sin tX + b\cos tY = (b^2 - a^2)\sin t\cos t,$$
$$-a\cos tX - b\sin tY = (b^2 - a^2)((\cos t)^2 - (\sin t)^2), \quad t \in [0, 2\pi],$$

or

$$-a\sin tX + b\cos tY = \frac{b^2 - a^2}{2}\sin(2t), \tag{2.5}$$
$$-a\cos tX - b\sin tY = (b^2 - a^2)\cos(2t), \quad t \in [0, 2\pi].$$

Multiplying the first equation of the latter system by $\sin t$ and the second by $\cos t$, we get

$$-aX = (b^2 - a^2)((\sin t)^2 \cos t + (\cos t)^3 - (\sin t)^2 \cos t)$$
$$= (b^2 - a^2)(\cos t)^3, \quad t \in [0, 2\pi],$$

whereupon

$$X = \frac{a^2 - b^2}{a}(\cos t)^3, \quad t \in [0, 2\pi].$$

Now, we multiply the first equation of the system (2.5) by $\cos t$ and the second by $-\sin t$ and find

$$bY = (b^2 - a^2)(\sin t(\cos t)^2 - \sin t(\cos t)^2 + (\sin t)^3)$$
$$= (b^2 - a^2)(\sin t)^3, \quad t \in [0, 2\pi],$$

or

$$Y = \frac{b^2 - a^2}{b}(\sin t)^3, \quad t \in [0, 2\pi].$$

Thus, the evolute of the ellipse is

$$\left(\frac{a^2 - b^2}{a}(\cos t)^3, \frac{b^2 - a^2}{b}(\sin t)^3 \right), \quad t \in [0, 2\pi].$$

Exercise 2.5. Find the evolutes of the following curves:
1.
$$f(t) = (t - \sin t, 1 - \cos t), \quad t \in [0, 2\pi];$$

2.
$$y^2 = 2px, \quad (x, y) \in \mathbb{R}^2,$$

where $p > 0$ is a given parameter;
3.
$$f(t) = (a \cosh t, b \sinh t), \quad t \in \mathbb{R},$$

where $a, b \in \mathbb{R}, (a, b) \neq (0, 0)$;
4.
$$f(t) = (a(\cos t)^3, a(\sin t)^3), \quad t \in [0, 2\pi],$$

where $a \in \mathbb{R}$;
5.
$$y = x^{2k}, \quad k \in \mathbb{N}, \quad (x, y) \in \mathbb{R}^2.$$

Answer 2.3. 1.
$$(t + \sin t, -1 + \cos t), \quad t \in [0, 2\pi];$$

2.
$$27py^2 = 8(x - p)^3, \quad (x, y) \in \mathbb{R}^2;$$

3.
$$(ax)^{\frac{2}{3}} + (by)^{\frac{2}{3}} = (a^2 - b^2)^{\frac{2}{3}}, \quad (x,y) \in \mathbb{R}^2;$$

4.
$$(x+y)^{\frac{2}{3}} + (x-y)^{\frac{2}{3}} = 2a^{\frac{2}{3}}, \quad (x,y) \in \mathbb{R}^2;$$

5.
$$\left(\frac{1}{2k-1}((2k-2)t - 4k^2t^{4k-1}), \frac{1+2k(4k-1)t^{4k-2}}{2k(2k-1)t^{2k-2}} \right), \quad t \in \mathbb{R}.$$

2.3 The complex structure on $\mathbb{R}^2$

Definition 2.3. The complex structure on $\mathbb{R}^2$ is the map $J : \mathbb{R}^2 \to \mathbb{R}^2$ defined by

$$Ju = (-u_2, u_1), \quad u = (u_1, u_2) \in \mathbb{R}^2.$$

Suppose that $\{e_1, e_2, e_3\}$ is an orthonormal basis in $\mathbb{R}^3$ and

$$u = (u_1, u_2), \quad v = (v_1, v_2) \in \mathbb{R}^2.$$

Exercise 2.6. Prove that

$$Ju \cdot Jv = u \cdot v.$$

Solution. We have

$$Ju = (-u_2, u_1),$$
$$Jv = (-v_2, v_1).$$

Then

$$Ju \cdot Jv = (-u_2)(-v_2) + u_1 v_1$$
$$= u \cdot v.$$

This completes the solution.

Exercise 2.7. Prove that

$$Ju \cdot u = 0.$$

Solution. We have

$$Ju = (-u_2, u_1)$$

and

$$Ju \cdot u = (-u_2)(u_1) + u_1 u_2 = 0.$$

This completes the solution.

Exercise 2.8. Prove that

$$v \cdot Ju = (u \times v) \cdot e_3.$$

Solution. We have

$$u \times v = \begin{vmatrix} e_1 & e_2 & e_3 \\ u_1 & u_2 & 0 \\ v_1 & v_2 & 0 \end{vmatrix}$$

$$= \begin{vmatrix} u_1 & u_2 \\ v_1 & v_2 \end{vmatrix} e_3$$

$$= (u_1 v_2 - u_2 v_1) e_3$$

and

$$Ju = (-u_2, u_1),$$
$$(u \times v) \cdot e_3 = u_1 v_2 - u_2 v_1,$$

as well as

$$v \cdot Ju = (-u_2)v_1 + u_1 v_2$$
$$= u_1 v_2 - u_2 v_1.$$

Thus, we get the desired result. This completes the proof.

Exercise 2.9. Show that

$$J(Ju) = -u.$$

Solution. We have

$$Ju = (-u_2, u_1)$$

and

$$J(Ju) = (-u_1, -u_2)$$
$$= (-u_1, -u_2)$$
$$= -(u_1, u_2)$$
$$= -u.$$

This completes the solution.

2.4 Curvature of plane curves

Suppose that $(I, f = f(t))$ is a parameterized plane curve.

Definition 2.4. The signed curvature of f is defined by

$$\kappa_{\pm} = \frac{f'' \cdot Jf'}{|f'|^3}.$$

Let

$$f = (f_1, f_2).$$

Then

$$f' = (f_1', f_2'),$$
$$f'' = (f_1'', f_2''),$$
$$Jf' = (-f_2', f_1').$$

Example 2.6. Let $f : \mathbb{R} \to \mathbb{R}^2$ be defined by

$$f(t) = (t^2, t + t^2), \quad t \in \mathbb{R}.$$

Here

$$f_1(t) = t^2,$$
$$f_2(t) = t + t^2, \quad t \in \mathbb{R}.$$

Then

$$f_1'(t) = 2t,$$
$$f_2'(t) = 1 + 2t, \quad t \in \mathbb{R},$$

and

$$f_1''(t) = 2,$$
$$f_2''(t) = 2, \quad t \in \mathbb{R}.$$

Then

$$f'(t) = (f_1'(t), f_2'(t))$$
$$= (2t, 1 + 2t),$$
$$Jf'(t) = (-1 - 2t, 2t),$$

2.4 Curvature of plane curves

$$f''(t) = (f_1''(t), f_2''(t))$$
$$= (2, 2), \quad t \in \mathbb{R},$$

and

$$f''(t) \cdot Jf'(t) = 2 \cdot (-1 - 2t) + 2 \cdot 2t$$
$$= -2 - 4t + 4t$$
$$= -2, \quad t \in \mathbb{R},$$

as well as

$$|f'(t)|^3 = \left((f_1'(t))^2 + (f_2'(t))^2 \right)^{\frac{3}{2}}$$
$$= \left((2t)^2 + (1 + 2t)^2 \right)^{\frac{3}{2}}$$
$$= \left(4t^2 + 1 + 4t + 4t^2 \right)^{\frac{3}{2}}$$
$$= (8t^2 + 4t + 1)^{\frac{3}{2}}, \quad t \in \mathbb{R}.$$

Therefore the signed curvature of the considered curve is

$$\kappa_\pm(t) = -\frac{2}{(8t^2 + 4t + 1)^{\frac{3}{2}}}, \quad t \in \mathbb{R}.$$

Example 2.7. We will find the signed curvature of the following curve:

$$y = \sin x, \quad x \in [0, 2\pi].$$

Here

$$f(x) = (x, \sin x),$$
$$f_1(x) = x,$$
$$f_2(x) = \sin x, \quad x \in [0, 2\pi].$$

Then

$$f_1'(x) = 1,$$
$$f_2'(x) = \cos x,$$
$$f_1''(x) = 0,$$
$$f_2''(x) = -\sin x, \quad x \in [0, 2\pi],$$

and

$$f'(x) = (f_1'(x), f_2'(x))$$
$$= (1, \cos x),$$
$$Jf'(x) = (-\cos x, 1),$$
$$f''(x) = (f_1''(x), f_2''(x))$$
$$= (0, -\sin x), \quad x \in [0, 2\pi].$$

Hence,

$$f''(x) \cdot Jf'(x) = -\sin x,$$
$$|f'(x)|^3 = ((f'(x))^2 + (f_2'(x))^2)^{\frac{3}{2}}$$
$$= (1 + (\cos x)^2)^{\frac{3}{2}}, \quad x \in [0, 2\pi].$$

Therefore the curvature of the considered curve is

$$\kappa_{\pm} = -\frac{\sin x}{(1 + (\cos x)^2)^{\frac{3}{2}}}, \quad x \in [0, 2\pi].$$

Exercise 2.10. Find the signed curvature of the following curves:

1.
$$y = a \cosh\left(\frac{x}{a}\right), \quad x \in \mathbb{R};$$

2.
$$y^2 = 2px, \quad x \in \mathbb{R},$$

where $p \in \mathbb{R}, p > 0$, is a parameter; .

3.
$$f(t) = (t^2, t^3), \quad t \in \mathbb{R};$$

4.
$$f(t) = (a \cos t, b \sin t), \quad t \in [0, 2\pi],$$

where $a, b \in \mathbb{R}$ are parameters;

5.
$$f(t) = (a \cosh t, b \sinh t), \quad t \in \mathbb{R},$$

where $a, b \in \mathbb{R}$ are parameters.

Answer 2.4. 1.
$$\frac{1}{ay^2}, \quad (x, y) \in \mathbb{R}^2;$$

2.
$$\frac{-\sqrt{p}}{(p + 2x)^{\frac{3}{2}}}, \quad \frac{-p^2}{(y^2 + p^2)^{\frac{3}{2}}}, \quad (x, y) \in \mathbb{R}^2;$$

3.
$$\frac{6}{t(4 + 9t^2)^{\frac{3}{2}}}, \quad t \in \mathbb{R};$$

4.
$$\frac{ab}{(a^2(\sin t)^2 + b^2(\cos t)^2)^{\frac{3}{2}}}, \quad t \in [0, 2\pi];$$

5.
$$\frac{-ab}{(a^2(\sinh t)^2 + b^2(\cosh t)^2)^{\frac{3}{2}}}, \quad t \in [0, 2\pi];$$

Note that

$$\kappa_{\pm} = \frac{f' \times f''}{|f'|^3}.$$

Thus,

$$|\kappa_{\pm}| = \kappa.$$

Exercise 2.11. Let $(I, f = f(t))$ be a parameterized curve and $(J, g = g(s(t)))$ be a naturally parameterized curve which is equivalent to f. Prove that

$$\kappa_{\pm}^g(s(t)) = \kappa_{\pm}^f(t), \quad t \in I.$$

Solution. We have

$$f(t) = g(s(t)), \quad t \in I,$$

and

$$f'(t) = g'(s(t))s'(t),$$
$$f''(t) = g''(s(t))(s'(t))^2 + g'(s(t))s''(t), \quad t \in I.$$

Hence,

$$(f''(t)) \cdot (Jf'(t)) = (g''(s(t))(s'(t))^2 + g'(s(t))s''(t)) \cdot (Jg'(s(t))s'(t))$$
$$= (g''(s(t))(s'(t))^2) \cdot (s'(t)Jg'(s(t)))$$
$$\quad + (g'(s(t))s''(t)) \cdot (s'(t)Jg'(s(t)))$$
$$= (s'(t))^3(g''(s(t)) \cdot Jg'(s(t))) + (s''(t)s'(t))(g'(s(t)) \cdot Jg'(s(t)))$$
$$= (s'(t))^3(g''(s(t)) \cdot Jg'(s(t)))$$
$$= (s'(t))^3 \kappa_{\pm}^g(s(t))$$
$$= |f'(t)|^3 \kappa_{\pm}^g(s(t)), \quad t \in I,$$

whereupon

$$\kappa_{\pm}^{g}(s(t)) = \frac{f''(t) \cdot Jf'(t)}{|f'(t)|^3}$$
$$= \kappa_{\pm}^{f}(t), \quad t \in I.$$

This completes the solution.

Exercise 2.12. Let $(I, f = f(s))$ be a naturally parameterized curve. Prove that

$$f''(s) = \kappa_{\pm}(s)Jf'(s), \quad s \in I.$$

Solution. We have

$$f'(s) \cdot f'(s) = 1, \quad s \in I.$$

Then

$$f'(s) \cdot f''(s) = 0, \quad s \in I,$$

and

$$f' \perp f'' \quad \text{on } I.$$

Since

$$f' \perp Jf' \quad \text{on } I,$$

we get

$$f'' \parallel Jf' \quad \text{on } I,$$

and there is a function a on I such that

$$f''(s) = a(s)Jf'(s), \quad s \in I.$$

Hence,

$$\kappa_{\pm}(s) = f''(s) \cdot Jf'(s)$$
$$= a(s)(Jf'(s) \cdot Jf'(s))$$
$$= a(s)(f'(s) \cdot f'(s))$$
$$= a(s), \quad s \in I.$$

Consequently,

$$f''(s) = \kappa_\pm(s)Jf'(s), \quad s \in I.$$

This completes the proof.

2.5 Rotation angle of plane curves

Let $(I, f = f(t))$ be a parameterized plane curve.

Definition 2.5. The rotation angle of f is the function $\theta : I \to \mathbb{R}$ defined by

$$\mathbf{t}(t) = (\cos \theta(t), \sin \theta(t)), \quad t \in I.$$

Here $\mathbf{t}$ is the unit tangent vector to f.

Exercise 2.13. Prove that

$$\theta'(t) = \kappa_\pm(t)|f'(t)|, \quad t \in I.$$

Solution. We have

$$\mathbf{t}(t) = \frac{f'(t)}{|f'(t)|}, \quad t \in I.$$

Then

$$\mathbf{t}'(t) = \frac{1}{|f'(t)|}f''(t) + \left(\frac{1}{|f'(t)|}\right)' f'(t), \quad t \in I,$$

and

$$\begin{aligned}
\mathbf{t}'(t) &= (-\sin\theta(t)\theta'(t), \cos\theta(t)\theta'(t)) \\
&= \theta'(t)J\mathbf{t}(t) \\
&= \theta'(t)J\left(\frac{f'(t)}{|f'(t)|}\right) \\
&= \theta'(t)J\left(\frac{f'(t)}{|Jf'(t)|}\right), \quad t \in I.
\end{aligned}$$

Thus,

$$\frac{1}{|f'(t)|}f''(t) + \left(\frac{1}{|f'(t)|}\right)' f'(t) = \theta'(t)J\left(\frac{f'(t)}{|Jf'(t)|}\right), \quad t \in I.$$

Hence,

$$(f''(t)(1/|f'(t)|)) \cdot Jf'(t) + \left(f'(t)\left(\frac{1}{|f'(t)|}\right)'\right) \cdot Jf'(t)$$
$$= \theta'(t)\frac{1}{|Jf'(t)|}(Jf'(t) \cdot Jf'(t)), \quad t \in I,$$

or

$$\left(\frac{1}{|f'(t)|}\right)(f''(t) \cdot Jf'(t)) = \theta'(t)|Jf'(t)|$$
$$= \theta'(t)|f'(t)|, \quad t \in I,$$

whereupon

$$\theta'(t) = (f''(t) \cdot Jf'(t))(|f'(t)|^2)$$
$$= \kappa_{\pm}(t)|f'(t)|, \quad t \in I.$$

This completes the proof.

Example 2.8. Consider the curve in Example 2.6. Using the computations in Example 2.6, we find

$$\theta(t) = -\arctan(4t+1) + a, \quad t \in \mathbb{R},$$

for some constant a.

Example 2.9. Consider the curve in Example 2.7. Using the computations in Example 2.7, we find

$$\theta(x) = -\arctan(\cos x) + a, \quad x \in \mathbb{R},$$

for some constant a.

Exercise 2.14. Find derivatives of the rotation angle of the curves in Exercise 2.10.

Answer 2.5. 1.
$$\frac{a}{y}, \quad (x,y) \in \mathbb{R}^2;$$

2.
$$\frac{1}{p+2x}, \quad \frac{p}{y^2+p^2}, \quad (x,y) \in \mathbb{R}^2;$$

3.
$$\frac{6}{4+9t^2}, \quad t \in \mathbb{R};$$

4.
$$\frac{ab}{a^2(\sin t)^2 + b^2(\cos t)^2}, \quad t \in [0, 2\pi];$$

5.
$$\frac{ab}{a^2(\sinh t)^2 + b^2(\cosh t)^2}, \quad t \in [0, 2\pi].$$

2.6 The curvature center

Definition 2.6. Suppose that $f : I \to \mathbb{R}^2$ is a parameterized plane curve. A point $g \in \mathbb{R}^2$ is said to be the curvature center at $f_0 = f(t_0)$, $t_0 \in I$, of the plane curve f if there is a circle γ, centered at g, which is tangent to the curve f at t_0, such that the signed curvatures of f and γ at t_0 coincide. Thus, the position of the point g for arbitrary $t \in I$ is given by

$$g(t) = f(t) + \frac{1}{\kappa_{\pm}(t)} \frac{Jf'(t)}{|f'(t)|}.$$

Example 2.10. Let f be the curve as in Example 2.6. Using the computations in Example 2.6, we get

$$g(t) = (t^2, t + t^2) - \frac{(8t^2 + 4t + 1)^{\frac{3}{2}}}{2} \frac{1}{(8t^2 + 4t + 1)^{\frac{1}{2}}} (-1 - 2t, 2t)$$

$$= (t^2, t + t^2) - \frac{8t^2 + 4t + 1}{2}(-1 - 2t, 2t)$$

$$= \left(t^2 + \frac{8t^2 + 4t + 1}{2} + t(8t^2 + 4t + 1), t + t^2 - 8t^3 - 4t^2 - t \right)$$

$$= \left(\frac{2t^2 + 8t^2 + 4t + 1 + 16t^3 + 8t^2 + 2t}{2}, -7t^2 - 3t - 1 \right)$$

$$= \left(\frac{16t^3 + 18t^2 + 6t + 1}{2}, -8t^3 - 3t^2 \right), \quad t \in \mathbb{R}.$$

Example 2.11. Let f be the curve as in Example 2.7. Using the computations in Example 2.7, we get

$$g(x) = (x, \sin x) - \frac{(1 + (\cos x)^2)^{\frac{1}{2}}}{\sin x} \frac{1}{(1 + (\cos x)^2)^{\frac{1}{2}}} (-\cos x, 1)$$

$$= (x, \sin x) - \frac{1 + (\cos x)^2}{\sin x}(-\cos x, 1)$$

$$= \left(x + \frac{\cos x + (\cos x)^3}{\sin x}, \sin x - \frac{1 + (\cos x)^2}{\sin x} \right)$$

$$= \left(\frac{x \sin x + \cos x + (\cos x)^3}{\sin x}, \frac{(\sin x)^2 - 1 - (\cos x)^2}{\sin x} \right)$$

$$= \left(\frac{x \sin x + \cos x + (\cos x)^3}{\sin x}, \frac{-1 - \cos(2x)}{\sin x} \right), \quad x \in [0, 2\pi].$$

Exercise 2.15. Find the curvature center of the curves in Exercise 2.10.

Answer 2.6. 1.

$$\left(x - a \sinh\left(\frac{2x}{a} \right), 2a \cosh\left(\frac{2x}{a} \right) \right), \quad x \in \mathbb{R};$$

2.

$$\left(-\frac{y^2 + 2p^2}{2p}, -\frac{y^3}{p^2}\right), \quad y \in \mathbb{R};$$

3.

$$\left(-\frac{2t^2 + 9t^4}{2}, \frac{4t + 12t^3}{3}\right), \quad t \in \mathbb{R};$$

4.

$$\left(\frac{a^2 - b^2}{a}(\cos t)^3, \frac{b^2 - a^2}{b}(\sin t)^3\right), \quad t \in [0, 2\pi];$$

5.

$$\left(\frac{a^2 + b^2}{a}(\cosh t)^3, -\frac{a^2 + b^2}{b}(\sinh t)^3\right), \quad t \in \mathbb{R}.$$

2.7 The involute

Definition 2.7. Let $f : I \to \mathbb{R}^2$ be a naturally parameterized curve and $c \in I$. The involute of f with origin at $f(c)$ is the parameterized curve

$$g(s) = f(s) + (c - s)f'(s), \quad s \in I.$$

If $f : I \to \mathbb{R}^2$ is an arbitrary parameterized curve, then we can replace the parameter t by the arc length

$$s = \int_0^t |f'(u)| du$$

and define the involute of f as being the involute of the naturally parameterized curve equivalent to it. In this case, the involute is given by the equation

$$g(t) = f(t) + (c - s(t))\frac{f'(t)}{|f'(t)|}, \quad t \in I,$$

where $s = s(t)$ is the arc length of f.

Example 2.12. Consider the circle

$$f(t) = (\cos t, \sin t), \quad t \in [0, 2\pi].$$

We have that $f : [0, 2\pi] \to \mathbb{R}^2$ is a naturally parameterized curve and, for any $c \in [0, 2\pi]$, the equation of the involute of the considered circle is (see Fig. 2.4 for $c = 0$)

$$\begin{aligned} g(t) &= f(t) + (c - t)f'(t) \\ &= (\cos t, \sin t) + (c - t)(-\sin t, \cos t) \\ &= (\cos t - (c - t)\sin t, \sin t + (c - t)\cos t), \quad t \in [0, 2\pi]. \end{aligned}$$

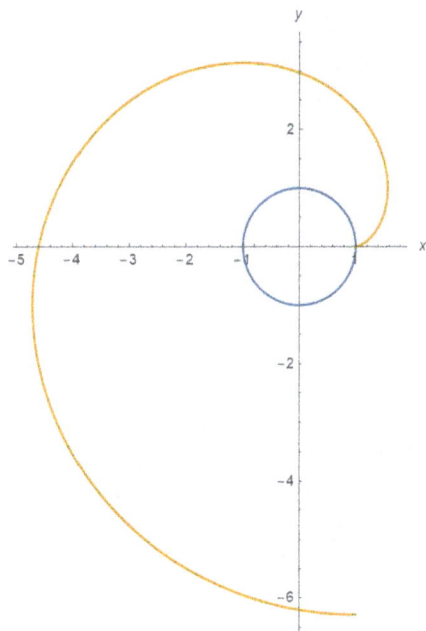

Figure 2.4: The involute of a unit circle given by $(\cos t + t \sin t, \sin t - t \cos t)$, $t \in [0, 2\pi]$: (blue) the circle; (orange) the involute.

Exercise 2.16. Find the involute of the curve

$$y = a \cosh \frac{x}{a}, \quad x \in \mathbb{R},$$

where $a > 0$ is a given parameter.

Answer 2.7.

$$\left(t + \frac{c - \sinh(\frac{x}{a})}{\cosh(\frac{x}{a})}, a \cosh\left(\frac{x}{a}\right) + \left(c - \sinh\left(\frac{x}{a}\right)\right) \tanh\left(\frac{x}{a}\right)\right), \quad t \in \mathbb{R}.$$

Exercise 2.17. Let $f : I \to \mathbb{R}^2$ be a naturally parameterized curve and g be the involute of f with the origin at $c \in I$. Prove that

$$\kappa_\pm^g(s) = \frac{\mathrm{sign}(\kappa_\pm^f(s))}{|c - s|}, \quad s \in I.$$

Solution. By the definition of involute, we have

$$g(s) = f(s) + (c - s)f'(s), \quad s \in I,$$

whereupon, after differentiating with respect to s, we get

$$g'(s) = f'(s) - f'(s) + (c - s)f''(s)$$
$$= (c - s)f''(s), \quad s \in I.$$

By the definition for curvature, we have

$$f''(s) = \kappa^f_\pm(s)Jf'(s), \quad s \in I.$$

Hence,

$$g'(s) = (c - s)\kappa^f_\pm(s)Jf'(s), \quad s \in I,$$

and

$$g''(s) = -\kappa^f_\pm(s)Jf'(s) + (c - s)(\kappa^f_\pm)'(s)Jf'(s) + (c - s)\kappa^f_\pm(s)Jf''(s)$$
$$= (-\kappa^f_\pm(s) + (c - s)(\kappa^f_\pm)'(s))Jf'(s) + (c - s)\kappa^f_\pm(s)Jf''(s), \quad s \in I.$$

Note that

$$\kappa^f_\pm(s) = f''(s) \cdot Jf'(s), \quad s \in I,$$

from where

$$\kappa^f_\pm(s) = -Jf''(s) \cdot f'(s), \quad s \in I,$$

and

$$Jf''(s) = -\kappa^f_\pm(s)f'(s), \quad s \in I.$$

Then

$$g''(s) = (-\kappa^f_\pm(s) + (c - s)(\kappa^f_\pm)'(s))Jf'(s) - (c - s)(\kappa^f_\pm(s))^2 f'(s), \quad s \in I,$$

and

$$Jg'(s) = -(c - s)\kappa^f_\pm(s)f'(s),$$
$$|g'(s)| = |c - s||\kappa^f_\pm(s)|, \quad s \in I.$$

Therefore

$$\kappa^g_\pm(s) = \frac{g''(s) \cdot Jg'(s)}{|g'(s)|^3}$$

$$= \frac{1}{|c - s|^3|\kappa^f_\pm(s)|^3}((-\kappa^f_\pm(s) + (c - s)(\kappa^f_\pm)'(s))Jf'(s) - (c - s)(\kappa^f_\pm(s))^2 f'(s))$$
$$\cdot (-(c - s)\kappa^f_\pm(s)f'(s))$$

$$= \frac{1}{|c-s|^3|\kappa_\pm^f(s)|^3}|c-s|^2(\kappa_\pm^f(s))^3$$

$$= \frac{\mathrm{sign}(\kappa_\pm^f(s))}{|c-s|}, \quad s \in I.$$

This completes the proof.

Exercise 2.18. Let $f : I \to \mathbb{R}^2$ be a naturally parameterized curve and g be its involute with the origin at $c \in I$. Prove that the evolute of g is f.

Solution. The evolute of g is given by the equation

$$g_1(s) = g(s) + \frac{1}{\kappa_\pm^g(s)}\frac{Jg'(s)}{|g'(s)|}$$

$$= f(s) + (c-s)f'(s) + \frac{|c-s|}{\mathrm{sign}(\kappa_\pm^f(s))}\frac{-(c-s)\kappa_\pm^f(s)f'(s)}{|(c-s)\kappa_\pm^f(s)Jf'(s)|}$$

$$= f(s) + (c-s)f'(s) + \frac{|c-s|}{\mathrm{sign}(\kappa_\pm^f(s))}\frac{(c-s)\kappa_\pm^f(s)J^2f'(s)}{|c-s||\kappa_\pm^f(s)||Jf'(s)|}$$

$$= f(s) + (c-s)f'(s) + (c-s)J^2f'(s)$$

$$= f(s) + (c-s)f'(s) - (c-s)f'(s)$$

$$= f(s), \quad s \in I.$$

This completes the proof.

2.8 The osculating circle of a curve

Definition 2.8. Let $f : I \to \mathbb{R}^2$ be a parameterized curve. The osculating circle of f at a point $t \in I$ is the circle centered at the curvature center $g(t)$ with radius equal to $\frac{1}{\kappa_\pm(t)}$.

Exercise 2.19. Let $f : I \to \mathbb{R}^2$ be a plane parameterized curve and $t_1 < t_2 < t_3 \in I$, $C(t_1, t_2, t_3)$ be the circle passing through $f(t_1), f(t_2)$, and $f(t_3)$, and assume there is a $t \in I$ such that $\kappa_\pm(t) \neq 0$. Prove that the osculating circle of f at the point t is the circle

$$C(t) = \lim_{\substack{t_1 \to t \\ t_2 \to t \\ t_3 \to t}} C(t_1, t_2, t_3).$$

Solution. Let $A(t_1, t_2, t_3)$ be the center of the circle $C(t_1, t_2, t_3)$ and $h : I \to \mathbb{R}$ be the function defined by

$$h(t) = |f(t) - A(t_1, t_2, t_3)|^2, \quad t \in I.$$

Then h is a smooth function and

$$h'(t) = f'(t) \cdot (f(t) - A(t_1, t_2, t_3)) + (f(t) - A(t_1, t_2, t_3)) \cdot f'(t)$$
$$= 2f'(t) \cdot (f(t) - A(t_1, t_2, t_3)),$$
$$h''(t) = 2f''(t) \cdot (f(t) - A(t_1, t_2, t_3)) + 2f'(t) \cdot f'(t)$$
$$= 2f''(t) \cdot (f(t) - A(t_1, t_2, t_3)) + 2|f'(t)|^2, \quad t \in I.$$

(2.6)

Since h is differentiable and

$$h(t_1) = h(t_2) = h(t_3),$$

applying the mean value theorem, we get that there are points u_1, u_2 so that

$$t_1 < u_1 < t_2 < u_2 < t_3$$

and

$$h'(u_1) = h'(u_2) = 0.$$

Now, we apply again the mean value theorem, and we find that there is a point $v_1 \in (u_1, u_2)$ such that

$$h''(v_1) = 0.$$

By (2.6), we obtain

$$h'(u_1) = 2f'(u_1) \cdot (f(u_1) - A(t_1, t_2, t_3)),$$
$$h'(u_2) = 2f'(u_2) \cdot (f(u_2) - A(t_1, t_2, t_3)),$$
$$h''(v_1) = 2f''(v_1) \cdot (f(u_1) - A(t_1, t_2, t_3)) + 2|f'(v_1)|^2.$$

Hence, letting $t_1, t_2, t_3 \to t$, we arrive at

$$f'(t) \cdot (f(t) - A(t)) = 0,$$
$$f''(t) \cdot (f(t) - A(t)) = -|f'(t)|^2,$$

(2.7)

where

$$A(t) = \lim_{\substack{t_1 \to t \\ t_2 \to t \\ t_3 \to t}} A(t_1, t_2, t_3).$$

By the first equality of (2.7), we conclude

$$f(t) - A(t) = \frac{(f(t) - A(t)) \cdot Jf'(t)}{|f'(t)|^2} Jf'(t).$$

Considering in the second equality of (2.7), we obtain

$$(f''(t) \cdot Jf'(t))((f(t) - A(t)) \cdot Jf'(t)) = -|f'(t)|^4.$$

Here, because $|f'(t)|^2 = Jf'(t) \cdot Jf'(t)$,

$$(f''(t) \cdot Jf'(t))(f(t) - A(t)) = -|f'(t)|^2 Jf'(t),$$

whereupon

$$\frac{f''(t) \cdot Jf'(t)}{|f'(t)|^3}(f(t) - A(t)) = -\frac{1}{|f'(t)|}Jf'(t),$$

$$\kappa_\pm(t)(f(t) - A(t)) = -\frac{1}{|f'(t)|}Jf'(t),$$

and

$$f(t) - A(t) = -\frac{1}{\kappa_\pm(t)|f'(t)|}Jf'(t),$$

or

$$f(t) = A(t) - \frac{1}{\kappa_\pm(f)(t)|f'(t)|}Jf'(t).$$

This completes the proof.

Exercise 2.20. Let $m : I \to \mathbb{R}$ be a continuous function.
1. Prove that there is a regular naturally parameterized curve $g : I \to \mathbb{R}^2$ such that

$$\kappa_\pm(g)(s) = m(s), \quad s \in I.$$

2. Prove that g is unique up to a proper motion of $\mathbb{R}^2$.

Hint 2.8. Use the solution of Problem 1.18.

Example 2.13. Let $a \in \mathbb{R}$, $a \neq 0$, and $m : \mathbb{R} \to \mathbb{R}$ be a function defined by

$$m(t) = a, \quad t \in \mathbb{R}.$$

By Exercise 2.20, it follows that there exists a unique curve $g : \mathbb{R} \to \mathbb{R}^2$ such that

$$\kappa_\pm^g(s) = a, \quad s \in \mathbb{R}.$$

Let θ be the rotation angle of g. Then

$$\theta'(s) = a, \quad s \in \mathbb{R},$$

whereupon

$$\theta(s) = \alpha s + \theta_0, \quad s \in \mathbb{R},$$

where θ_0 is a constant. Let now $\mathbf{t}$ be the unit tangent vector of g. Then

$$\begin{aligned}
\mathbf{t}(s) &= (g_1'(s), g_2'(s)) \\
&= (\cos(\theta(s)), \sin(\theta(s))) \\
&= (\cos(\alpha s + \theta_0), \sin(\alpha s + \theta_0)), \quad s \in I.
\end{aligned}$$

Thus, we get the system

$$\begin{aligned}
g_1'(s) &= \cos(\alpha s + \theta_0), \\
g_2'(s) &= \sin(\alpha s + \theta_0), \quad s \in \mathbb{R}.
\end{aligned}$$

Hence,

$$\begin{aligned}
g_1(s) &= \frac{1}{\alpha} \sin(\alpha s + \theta_0) + g_{10}, \\
g_2(s) &= -\frac{1}{\alpha} \cos(\alpha s + \theta_0) + g_{20}, \quad s \in I,
\end{aligned}$$

where g_{10}, g_{20} are given constants, or

$$\begin{aligned}
g_1(s) &= \frac{1}{\alpha} \sin(\alpha s) \cos \theta_0 + \frac{1}{\alpha} \cos(\alpha s) \sin \theta_0 + g_{10}, \\
g_2(s) &= -\frac{1}{\alpha} \cos(\alpha s) \cos \theta_0 + \frac{1}{\alpha} \sin(\alpha s) \sin \theta_0 + g_{20}, \quad s \in I,
\end{aligned}$$

or

$$\begin{aligned}
\begin{pmatrix} g_1(s) \\ g_2(s) \end{pmatrix} &= \begin{pmatrix} \frac{1}{\alpha} \sin(\alpha s) \cos \theta_0 + \frac{1}{\alpha} \cos(\alpha s) \sin \theta_0 \\ -\frac{1}{\alpha} \cos(\alpha s) \cos \theta_0 + \frac{1}{\alpha} \sin(\alpha s) \sin \theta_0 \end{pmatrix} + \begin{pmatrix} g_{10} \\ g_{20} \end{pmatrix} \\
&= \begin{pmatrix} \cos(\frac{\pi}{2} - \theta_0) & \sin(\frac{\pi}{2} - \theta_0) \\ -\sin(\frac{\pi}{2} - \theta_0) & \cos(\frac{\pi}{2} - \theta_0) \end{pmatrix} \begin{pmatrix} \frac{1}{\alpha} \cos(\alpha s) \\ \frac{1}{\alpha} \sin(\alpha s) \end{pmatrix} + \begin{pmatrix} g_{10} \\ g_{20} \end{pmatrix}, \quad s \in \mathbb{R}.
\end{aligned}$$

We here conclude that any plane curve of constant signed curvature can be obtained by

$$\begin{aligned}
g_1(s) &= \frac{1}{\alpha} \cos(\alpha s), \\
g_2(s) &= -\frac{1}{\alpha} \sin(\alpha s), \quad s \in \mathbb{R}.
\end{aligned}$$

Here a rotation following by a translation, i. e., a rigid motion was applied. Thus, the only plane curve of constant positive curvature α is the circle of radius $\frac{1}{\alpha}$.

Definition 2.9. Natural equations of a curve are equations of the following form:

$$\kappa = \kappa_{\pm}(s),$$
$$F(\kappa, s) = 0,$$
$$\begin{cases} \kappa = \kappa_{\pm}(t), \\ s = s(t), \end{cases}$$

where s is the arc length of the curve.

Example 2.14. Let $I = (0, \infty)$ and $f : I \to \mathbb{R}^2, f(s) = (f_1(s), f_2(s)), s \in I$, be a curve for which

$$\kappa_{\pm}(s) = \frac{1}{as}, \quad s \in I,$$

where $a > 0$ is a given parameter. Let θ be the rotation angle of the curve f. Then

$$\theta'(s) = \frac{1}{as}, \quad s \in I,$$

whereupon

$$\theta(s) = \frac{1}{a} \log s + \frac{b}{a}, \quad s \in I, \quad b \in \mathbb{R},$$

and

$$s = e^{a\theta(s)-b},$$
$$ds = ae^{a\theta(s)-b} d\theta(s), \quad s \in I.$$

Then, if $\mathbf{t}$ is the unit normal vector to f, we get

$$\mathbf{t}(s) = (\cos \theta(s), \sin \theta(s))$$
$$= (f_1'(s), f_2'(s)), \quad s \in I.$$

Thus, we get the following system:

$$f_1'(s) = \cos \theta(s),$$
$$f_2'(s) = \sin \theta(s), \quad s \in I,$$

whereupon

$$f_1(s) = \int \cos \theta(s) ds$$
$$= a \int \cos \theta(s) e^{a\theta(s)-b} d\theta(s)$$

$$= \frac{ae^{a\theta(s)-b}}{a^2+1}(a\cos\theta(s)+\sin\theta(s)),$$

$$f_2(s) = \int \sin\theta(s)ds$$

$$= a\int \sin\theta(s)e^{a\theta(s)-b}d\theta(s)$$

$$= \frac{ae^{a\theta(s)-b}}{a^2+1}(a\sin\theta(s)-\cos\theta(s)), \quad s\in I.$$

Thus, the parametric equations of the considered curve are

$$f_1(s) = \frac{ae^{\theta(s)-b}}{a^2+1}(a\cos\theta(s)+\sin\theta(s)),$$

$$f_2(s) = \frac{ae^{a\theta(s)-b}}{a^2+1}(a\sin\theta(s)-\cos\theta(s)), \quad s\in I.$$

Exercise 2.21. Let $a > 0$ be a given parameter. Find the curves that are determined by the following natural equations:

1.
$$\kappa_{\pm}(s) = a, \quad s\in\mathbb{R};$$

2.
$$\frac{1}{\kappa_{\pm}(s)} = a(1+s^2), \quad s\in\mathbb{R};$$

3.
$$s^2 + \frac{1}{(\kappa_{\pm}(s))^2} = 16a^2, \quad s\in\mathbb{R};$$

Answer 2.9. 1.
$$x^2 + y^2 = \frac{1}{a^2}, \quad (x,y)\in\mathbb{R}^2;$$

2.
$$y = a\cosh\frac{x}{a}, \quad (x,y)\in\mathbb{R}^2;$$

3.
$$\frac{1}{4a}\left(\frac{1}{2}s\sqrt{16a^2-s^2} + \arcsin\left(\frac{s}{4a}\right), \frac{s^2}{2}\right), \quad t\in\mathbb{R}.$$

2.9 Advanced practical problems

Problem 2.1. Find the envelopes of the following families of curves:

1.
$$F(x,y,\lambda) = (x-\lambda)^2 + y^2 - a^2 = 0, \quad (x,y)\in\mathbb{R}^2, \quad \lambda\in\mathbb{R},$$

where $a > 0$ is a given constant;

2.
$$F(x,y,\lambda) = y^3 - (x - \lambda)^2 = 0, \quad (x,y) \in \mathbb{R}^2, \quad \lambda \in \mathbb{R};$$

3.
$$F(x,y,\lambda) = 3(y - \lambda)^2 - 2(x - \lambda)^3 = 0, \quad (x,y) \in \mathbb{R}^2, \quad \lambda \in \mathbb{R};$$

4.
$$F(x,y,\lambda) = (1 - \lambda^2)x + 2\lambda y - a = 0, \quad (x,y) \in \mathbb{R}^2, \quad \lambda \in \mathbb{R},$$

where $a > 0$ is a given constant;

5.
$$F(x,y,\lambda) = \lambda^2(x - a) - \lambda y - a = 0, \quad (x,y) \in \mathbb{R}^2, \quad \lambda \in \mathbb{R},$$

where $a > 0$ is a given constant.

Answer 2.10. 1.
$$y = \pm a, \quad (x,y) \in \mathbb{R}^2;$$

2.
$$y = 0, \quad (x,y) \in \mathbb{R}^2;$$

3.
$$x - y = 0, \quad (x,y) \in \mathbb{R}^2;$$

4.
$$x^2 + y^2 - ax = 0, \quad (x,y) \in \mathbb{R}^2;$$

5.
$$y^2 + 4a(x - a) = 0, \quad (x,y) \in \mathbb{R}^2.$$

Problem 2.2. Find the evolutes of the following curves:

1.
$$y = x^{2k+1}, \quad k \in \mathbb{N}, \quad (x,y) \in \mathbb{R}^2;$$

2.
$$y = \log x, \quad x \in \mathbb{R}, \quad x > 0;$$

3.
$$y = \sin x, \quad x \in \mathbb{R};$$

4.
$$y = \tan x, \quad x \in \left(-\frac{\pi}{2}, \frac{\pi}{2}\right);$$

5.
$$\left(a\left(\log\left(\tan\left(\frac{t}{2}\right)\right) + \cos t\right), a \sin t\right), \quad t \in \mathbb{R},$$

where $a \in \mathbb{R}$;

6.
$$r = (1 + \cos \phi), \quad \phi \in [0, 2\pi].$$

Answer 2.11. 1.

$$\left(t + \frac{t}{2k} + (2k+1)t^{4k-1} - (2k+1)t^{4k} + \frac{(2k+1)^2}{2k}t^{4k+1},\right.$$

$$\left.\frac{t + 2k(2k+1)t^{4k} + (2k+1)^2t^{4k+1}}{2k(2k+1)t^{2k}}\right), \quad t \in \mathbb{R};$$

2.

$$\left(2t + \frac{1}{t}, \log t - t^2 - 1\right), \quad t \in \mathbb{R}, \quad t > 0;$$

3.

$$\left(t + \cos t\frac{1 + (\cos t)^2}{\sin t}, -\frac{2(\cos t)^2}{\sin t}\right), \quad t \in \mathbb{R};$$

4.

$$\left(t - \frac{1 + (\cos t)^4}{(\cos t)^2 \sin(2t)}, \tan t + \frac{1 + (\cos t)^4}{\sin(2t)}\right), \quad t \in \left(-\frac{\pi}{2}, \frac{\pi}{2}\right);$$

5.

$$\left(a\log\left(\tan\left(\frac{t}{2}\right)\right), \frac{a}{\sin t}\right), \quad t \in \mathbb{R},$$

or

$$\left(t, a\cosh\left(\frac{ty}{a}\right)\right), \quad t \in \mathbb{R};$$

6.

$$\left(\frac{a}{3}(\cos\phi - 2(\cos\phi)^3 + 2), \frac{a}{3}\sin\phi(1 - \cos\phi)\right), \quad \phi \in [0, 2\pi].$$

Problem 2.3. Find the curvature of the following curves:

1.

$$f(t) = (a(t - \sin t), a(-\cos t)), \quad t \in \mathbb{R},$$

where $a \in \mathbb{R}$ is a given positive parameter;

2.

$$f(t) = (a(1+m)\cos(mt) - am\cos((1+m)t),$$
$$a(1+m)\sin(mt) - am\sin((1+m)t)), \quad t \in [0, 2\pi],$$

where $a, m \in \mathbb{R}$ are given positive parameters;

3.

$$f(t) = (a(\cos t)^3, a(\sin t)^3), \quad t \in [0, 2\pi],$$

where $a \in \mathbb{R}$ is a given positive parameter;

4.

$$f(t) = (a(\cos t + t\sin t), a(\sin t - t\cos t)), \quad t \in [0, 2\pi],$$

where $a \in \mathbb{R}$ is a given positive parameter;

5.
$$r = a\phi, \quad \phi \in [0, 2\pi],$$

where $a \in \mathbb{R}$ is a given positive parameter;

6.
$$r = ae^{h\phi}, \quad \phi \in [0, 2\pi],$$

where $a, h \in \mathbb{R}$ are given positive parameters;

7.
$$r = a(1 + \cos \phi), \quad \phi \in [0, 2\pi],$$

where $a \in \mathbb{R}$ is a given positive parameter;

8.
$$r^2 = a^2 \cos(2\phi), \quad \phi \in [0, 2\pi],$$

where $a \in \mathbb{R}$ is a given positive parameter;

9.
$$\frac{x^2}{a^2} + \frac{y^2}{b^2} = 1, \quad (x, y) \in \mathbb{R}^2,$$

where $a, b \in \mathbb{R}$ are given positive parameters;

10.
$$\frac{x^2}{a^2} - \frac{y^2}{b^2} = 1, \quad (x, y) \in \mathbb{R}^2,$$

where $a, b \in \mathbb{R}$ are given positive parameters.

Answer 2.12. 1.
$$\frac{1}{4a|\sin(\frac{t}{2})|}, \quad t \in [0, 2\pi];$$

2.
$$\frac{1 + 2m}{4am(m + 1)|\sin(\frac{t}{2})|}, \quad t \in [0, 2\pi];$$

3.
$$\frac{2}{3a|\sin(2t)|}, \quad t \in [0, 2\pi];$$

4.
$$\frac{1}{at}, \quad t \in \mathbb{R};$$

5.
$$\frac{2 + \phi^2}{a(1 + \phi^2)^{\frac{3}{2}}}, \quad \phi \in [0, 2\pi];$$

6.
$$\frac{1}{r\sqrt{1 + h^2}}, \quad \phi \in [0, 2\pi];$$

7.
$$\frac{3}{4a|\cos(\frac{\phi}{2})|}, \quad \phi \in [0, 2\pi];$$

8.
$$\frac{3r}{a^2}, \quad \phi \in [0, 2\pi];$$

9.
$$\frac{a^4 b^4}{(b^4 x^2 + a^4 y^2)^{\frac{3}{2}}}, \quad (x, y) \in \mathbb{R}^2;$$

10.
$$\frac{a^4 b^4}{(b^4 x^2 + a^4 y^2)^{\frac{3}{2}}}, \quad (x, y) \in \mathbb{R}^2.$$

Problem 2.4. Find the derivatives of rotation angle of the curves in Problem 2.3.

Answer 2.13. 1.
$$\frac{1}{2}, \quad t \in [0, 2\pi];$$

2.
$$\frac{1 + 2m}{2}, \quad t \in [0, 2\pi];$$

3.
$$1, \quad t \in [0, 2\pi];$$

4.
$$1, \quad t \in \mathbb{R};$$

5.
$$\frac{2 + \phi^2}{1 + \phi^2}, \quad \phi \in [0, 2\pi];$$

6.
$$1, \quad \phi \in [0, 2\pi];$$

7.
$$\frac{3}{2}, \quad \phi \in [0, 2\pi];$$

8.
$$3, \quad \phi \in [0, 2\pi];$$

9.
$$\frac{a^3 b^3}{b^4 x^2 + a^4 y^2}, \quad (x, y) \in \mathbb{R}^2;$$

10.
$$\frac{a^3 b^3}{b^4 x^2 + a^4 y^2}, \quad (x, y) \in \mathbb{R}^2.$$

Problem 2.5. Find the curvature center of the curves in Problem 2.3.

Answer 2.14. 1.
$$(a(t - 3\sin t), a(1 - 3\cos t)), \quad t \in [0, 2\pi];$$

2.
$$\left(\frac{a(1 + m)}{1 + 2m} \cos(mt) + \frac{am}{1 + 2m} \cos((1 + m)t), \right.$$
$$\left. \frac{a(1 + m)}{1 + 2m} \sin(mt) + \frac{am}{1 + 2m} \sin((1 + m)t) \right), \quad t \in [0, 2\pi];$$

3.
$$(a((\cos t)^2 - 3(\sin t)^2) \cos t, a((\sin t)^2 - 3(\cos t)^2) \sin t), \quad t \in [0, 2\pi];$$

4.
$$(a \cos t, a \sin t), \quad t \in [0, 2\pi];$$

5.
$$\left(-a\frac{1+\phi^2}{2+\phi^2} \sin \phi + \frac{a\phi}{2+\phi^2} \cos \phi,\right.$$
$$\left.\frac{a(1+\phi^2)}{2+\phi^2} \cos \phi + \frac{a\phi}{2+\phi^2} \sin \phi\right), \quad \phi \in [0, 2\pi];$$

6.
$$(-ahe^{h\phi} \sin \phi, ahe^{h\phi} \cos \phi), \quad \phi \in [0, 2\pi];$$

7.
$$\left(a\left(\cos \phi + (\cos \phi)^2 - \frac{2}{3} \sin \phi - \frac{2}{3} \sin(2\phi)\right),\right.$$
$$\left.a\left(\sin \phi + \frac{1}{2} \sin(2\phi) + \frac{2}{3} \cos \phi + \frac{1}{3} \cos(2\phi)\right)\right), \quad \phi \in [0, 2\pi];$$

8.
$$\left(\frac{2}{3}r \cos \phi + \frac{a^2 \sin(2\phi) \sin \phi}{3r}, \frac{4}{3}r \sin \phi - \frac{a^2 \sin(2\phi) \cos \phi}{3r}\right), \quad \phi \in [0, 2\pi];$$

9.
$$\left(\frac{(a^2 - b^2)(\cos \phi)^3}{a}, \frac{(b^2 - a^2)(\sin \phi)^3}{b}\right), \quad \phi \in [0, 2\pi];$$

10.
$$\left(\frac{(a^4b^2 - a^4y^2 - b^4x^2)x}{a^4b^2}, \frac{(a^2b^4 + b^4x^2 + a^4y^2)y}{a^2b^4}\right), \quad (x,y) \in \mathbb{R}^2.$$

Problem 2.6. Prove that the evolute of a plane curve is the geometrical locus of the curvature centers of the curve.

Hint 2.15. Use the definition of the evolute.

Problem 2.7. Find the involute of the following curve:
$$\left(t, \frac{1}{4}t^2\right), \quad t \in \mathbb{R}.$$

Answer 2.16.
$$\left(\frac{t}{2} + \frac{2}{\sqrt{t^2+4}}(c - \log(t + \sqrt{t^2+4})),\right.$$
$$\left.\frac{t}{\sqrt{t^2+4}}(c - \log(t + \sqrt{t^2+4}))\right), \quad t \in \mathbb{R}.$$

Problem 2.8. Find the curves that are determined by the following natural equations:
1.
$$\frac{s^2}{a^2} + \frac{1}{b^2(\kappa_\pm(s))^2} = 1, \quad s \in \mathbb{R},$$

where $a, b > 0$ are given parameters;

2.
$$\frac{1}{(\kappa_{\pm}(s))^2} = 2as, \quad s \in \mathbb{R},$$

where $a > 0$ is a given parameter;

3.
$$\frac{1}{(\kappa_{\pm}(s))^2} + a^2 = a^2 e^{-\frac{2s}{a}}, \quad s \in \mathbb{R},$$

where $a > 0$ is a given parameter.

Answer 2.17. 1.
$$s = a \sin t,$$
$$R = b \cos t,$$
$$\theta(t) = \frac{at}{b}, \quad t \in \mathbb{R};$$

2.
$$(a(\cos t + t \sin t), a(\sin t - t \cos t)), \quad t \in \mathbb{R};$$

3.
$$\left(a \cos t, a\left(\log\left(\frac{1 + \sin t}{\cos t} \right) - \sin t \right) \right), \quad t \in \mathbb{R}.$$

Problem 2.9. Find the natural equations of the following curves:

1.
$$y = x^{\frac{3}{2}}, \quad x > 0;$$

2.
$$(a(\cos t + t \sin t), a(\sin t - t \cos t)), \quad t \in \mathbb{R};$$

3.
$$\left(a\left(\log\left(\tan \frac{t}{2} \right) + \cos t \right), a \sin t \right), \quad t \in \mathbb{R};$$

4.
$$r = a(1 + \cos \phi), \quad \phi \in \mathbb{R}.$$

Answer 2.18. 1.
$$(27s + 8)^2 = \left(4 + \frac{324}{(\kappa_{\pm}(s))^2(27s + 8)^2} \right)^3, \quad s \in \mathbb{R};$$

2.
$$\left(\frac{1}{\kappa_{\pm}(s)} \right)^2 = 2as, \quad s \in \mathbb{R};$$

3.
$$\left(\frac{1}{\kappa_{\pm}(s)} \right)^2 + a^2 = a^2 e^{-\frac{2s}{a}}, \quad s \in \mathbb{R};$$

4.
$$x^2 + \frac{9}{(\kappa_\pm(s))^2} = 16a^2, \quad s \in \mathbb{R}.$$

Problem 2.10. Let $a > 0$ be a given parameter. Find the parametric equations of the curves for which:

1.
$$\frac{1}{(\kappa_\pm(s))^2}(\sin\theta(s))^3 = a, \quad s \in \mathbb{R};$$

2.
$$s = a\tan\theta(s), \quad s \in \mathbb{R};$$

3.
$$s = a\cos\theta(s), \quad s \in \mathbb{R}.$$

Answer 2.19. 1.
$$\left(-\frac{a}{2(\sin\theta(s))^2}, -a\cot\theta(s)\right), \quad s \in \mathbb{R};$$

2.
$$\left(a\log\left(\tan\left(\frac{\pi}{4} + \frac{\theta(s)}{2}\right)\right), \frac{a}{\cos\theta(s)}\right), \quad s \in \mathbb{R};$$

3.
$$\left(-\frac{a}{4}(1 - \cos(2\theta(s))), -\frac{a}{4}(2\theta(s) - \sin(2\theta(s)))\right), \quad s \in \mathbb{R}.$$

3 General theory of surfaces

In this chapter we first define parametrized surfaces and their equivalence under diffeomorphisms. Regular surfaces are also given and studied with examples. We revisit curves by means of parametrized surfaces. The equations of the tangent plane and the normal to a surface are derived in different forms: parametric, nonparametric, and implicit. Considering the notion of differentiability of a map between two surfaces, we introduce the shape operator of a surface and investigate its properties. In addition, we deal with intrinsic invariants of a surface (the first fundamental form and the Gauss curvature) and extrinsic invariants (the second fundamental form, the principal curvatures and the mean curvature). Finally, the well-known Joachimsthal and Meusnier theorems are proved.

3.1 Parameterized surfaces

Suppose that $U \subseteq \mathbb{R}^2$.

Definition 3.1. A regular parameterized surface in $\mathbb{R}^3$ is a smooth map $f : U \to \mathbb{R}^3$, $(t_1, t_2) \mapsto f(t_1, t_2)$, and

$$f_{t_1} \times f_{t_2} \neq 0 \quad \text{on } U. \tag{3.1}$$

Definition 3.2. The requirement (3.1) is called the regularity condition.

Example 3.1. Let $f : \mathbb{R}^2 \to \mathbb{R}^3$ be given by

$$f(t_1, t_2) = (t_1 + t_2, e^{t_1^2 + t_2}, e^{t_1 + t_2^2}), \quad (t_1, t_2) \in \mathbb{R}^2.$$

Here

$$f_1(t_1, t_2) = t_1 + t_2,$$
$$f_2(t_1, t_2) = e^{t_1^2 + t_2},$$
$$f_3(t_1, t_2) = e^{t_1 + t_2^2}, \quad (t_1, t_2) \in \mathbb{R}^2.$$

Then

$$f_{1t_1}(t_1, t_2) = 1,$$
$$f_{1t_2}(t_1, t_2) = 1,$$
$$f_{2t_1}(t_1, t_2) = 2t_1 e^{t_1^2 + t_2},$$
$$f_{2t_2}(t_1, t_2) = e^{t_1^2 + t_2},$$
$$f_{3t_1}(t_1, t_2) = e^{t_1 + t_2^2},$$

https://doi.org/10.1515/9783111501857-003

$$f_{3t_2}(t_1, t_2) = 2t_2 e^{t_1 + t_2^2}, \quad (t_1, t_2) \in \mathbb{R}^2.$$

Hence,

$$f_{t_1}(t_1, t_2) = (f_{1t_1}(t_1, t_2), f_{2t_1}(t_1, t_2), f_{3t_1}(t_1, t_2))$$
$$= (1, 2t_1 e^{t_1^2 + t_2}, e^{t_1 + t_2^2}),$$
$$f_{t_2}(t_1, t_2) = (f_{1t_2}(t_1, t_2), f_{2t_2}(t_1, t_2), f_{3t_2}(t_1, t_2))$$
$$= (1, e^{t_1^2 + t_2}, 2t_2 e^{t_1 + t_2^2}), \quad (t_1, t_2) \in \mathbb{R}^2,$$

and

$$f_{t_1}(t_1, t_2) \times f_{t_2}(t_1, t_2) = ((4t_1 t_2 - 1)e^{t_1 + t_1^2 + t_2 + t_2^2},$$
$$(1 - 2t_2)e^{t_1 + t_2^2}, (1 - 2t_1)e^{t_1^2 + t_2}), \quad (t_1, t_2) \in \mathbb{R}^2.$$

Then

$$f_{t_1}(t_1, t_2) \times f_{t_2}(t_1, t_2) \neq 0$$

if and only if $t_1 \neq \frac{1}{2}$ or $t_2 \neq \frac{1}{2}$. Thus, the considered surface is regular for those $(t_1, t_2) \in \mathbb{R}^2$ with $t_1 \neq \frac{1}{2}$ or $t_2 \neq \frac{1}{2}$; see Fig. 3.1.

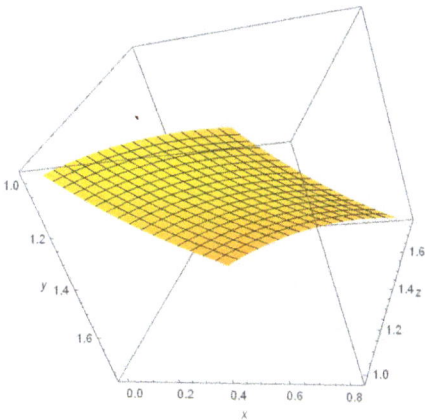

Figure 3.1: A regular parameterized surface $f(t_1, t_2) = (t_1 + t_2, e^{t_1^2 + t_2}, e^{t_1 + t_2^2})$, $(t_1, t_2) \in [0, 2/5]$.

Example 3.2. Consider the unit sphere (see Fig. 3.2)

$$f(t_1, t_2) = (\cos t_1 \cos t_2, \sin t_1 \cos t_2, \sin t_2), \quad t_1 \in (0, 2\pi), \quad t_2 \in \left(-\frac{\pi}{2}, \frac{\pi}{2}\right).$$

Here

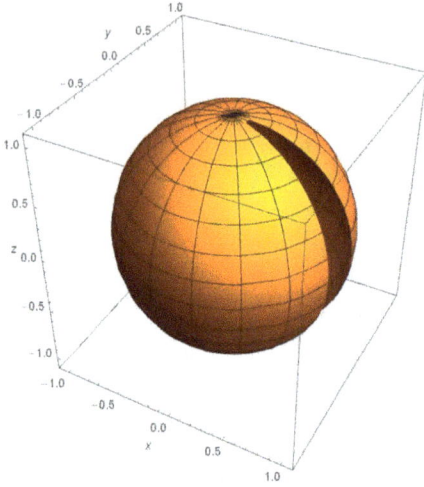

Figure 3.2: The unit sphere $f(t_1, t_2) = (\cos t_1 \cos t_2, \sin t_1 \cos t_2, \sin t_2)$, $t_1 \in (1/10, 6)$, $t_2 \in (-17/10, 17/10)$.

$$f_1(t_1, t_2) = \cos t_1 \cos t_2,$$
$$f_2(t_1, t_2) = \sin t_1 \cos t_2,$$
$$f_3(t_1, t_2) = \sin t_2, \quad t_1 \in (0, 2\pi), \quad t_2 \in \left(-\frac{\pi}{2}, \frac{\pi}{2}\right).$$

Then

$$f_{1t_1}(t_1, t_2) = -\sin t_1 \cos t_2,$$
$$f_{1t_2}(t_1, t_2) = -\cos t_1 \sin t_2,$$
$$f_{2t_1}(t_1, t_2) = \cos t_1 \cos t_2,$$
$$f_{2t_2}(t_1, t_2) = -\sin t_1 \sin t_2,$$
$$f_{3t_1}(t_1, t_2) = 0,$$
$$f_{3t_2}(t_1, t_2) = \cos t_2, \quad t_1 \in (0, 2\pi), \quad t_2 \in \left(-\frac{\pi}{2}, \frac{\pi}{2}\right),$$

and

$$f_{t_1}(t_1, t_2) = (f_{1t_1}(t_1, t_2), f_{2t_1}(t_1, t_2), f_{3t_1}(t_1, t_2))$$
$$= (-\sin t_1 \cos t_2, \cos t_1 \cos t_2, 0),$$
$$f_{t_2}(t_1, t_2) = (f_{1t_2}(t_1, t_2), f_{2t_2}(t_1, t_2), f_{3t_2}(t_1, t_2))$$
$$= (-\cos t_1 \sin t_2, -\sin t_1 \sin t_2, \cos t_2), \quad t_1 \in (0, 2\pi), \quad t_2 \in \left(-\frac{\pi}{2}, \frac{\pi}{2}\right).$$

Hence,

$$f_{t_1}(t_1, t_2) \times f_{t_2}(t_1, t_2)$$
$$= (\cos t_1(\cos t_2)^2, \sin t_1(\cos t_2)^2, (\sin t_1)^2 \sin t_2 \cos t_2 + (\cos t_1)^2 \sin t_2 \cos t_2)$$
$$= (\cos t_1(\cos t_2)^2, \sin t_1(\cos t_2)^2, \sin t_2 \cos t_2)$$
$$\neq 0, \quad t_1 \in (0, 2\pi), \quad t_2 \in \left(-\frac{\pi}{2}, \frac{\pi}{2}\right).$$

Thus, the unit sphere is a regular surface in $(0, 2\pi) \times (-\frac{\pi}{2}, \frac{\pi}{2})$.

Example 3.3. Consider the torus

$$f(t_1, t_2) = ((a + b\cos t_1)\cos t_2, (a + b\cos t_1)\sin t_2, b\sin t_1), \quad (t_1, t_2) \in [0, 2\pi] \times [0, 2\pi],$$

where $a, b \in \mathbb{R}$, $a > 2$, $b \in (0, 1)$. Here

$$f_1(t_1, t_2) = (a + b\cos t_1)\cos t_2,$$
$$f_2(t_1, t_2) = (a + b\cos t_1)\sin t_2,$$
$$f_3(t_1, t_2) = b\sin t_1, \quad (t_1, t_2) \in [0, 2\pi] \times [0, 2\pi].$$

Then

$$f_{1t_1}(t_1, t_2) = -b\sin t_1 \cos t_2,$$
$$f_{1t_2}(t_1, t_2) = -(a + b\cos t_1)\sin t_2,$$
$$f_{2t_1}(t_1, t_2) = -b\sin t_1 \sin t_2,$$
$$f_{2t_2}(t_1, t_2) = (a + b\cos t_1)\cos t_2,$$
$$f_{3t_1}(t_1, t_2) = b\cos t_1,$$
$$f_{3t_2}(t_1, t_2) = 0, \quad (t_1, t_2) \in [0, 2\pi] \times [0, 2\pi],$$

and

$$f_{t_1}(t_1, t_2) = (f_{1t_1}(t_1, t_2), f_{2t_1}(t_1, t_2), f_{3t_1}(t_1, t_2))$$
$$= (-b\sin t_1 \cos t_2, -b\sin t_1 \sin t_2, b\cos t_1),$$
$$f_{t_2}(t_1, t_2) = (f_{2t_1}(t_1, t_2), f_{2t_2}(t_1, t_2), f_{3t_2}(t_1, t_2))$$
$$= (-(a + b\cos t_1)\sin t_2, (a + b\cos t_1)\cos t_2, 0), \quad (t_1, t_2) \in [0, 2\pi] \times [0, 2\pi].$$

Hence,

$$f_{t_1}(t_1, t_2) \times f_{t_2}(t_1, t_2) = (-b(a + b\cos t_1)\cos t_2 \cos t_1, -b(a + b\cos t_1)\sin t_2 \cos t_1,$$
$$- b(a + b\cos t_1)\sin t_1(\cos t_2)^2 - b(a + b\cos t_1)\sin t_1(\sin t_2)^2)$$
$$= (-b(a + b\cos t_1)\cos t_1 \cos t_2, -b(a + b\cos t_1)\cos t_1 \sin t_2,$$
$$- b(a + b\cos t_1)\sin t_1)$$

or

$$f_{t_1}(t_1, t_2) \times f_{t_2}(t_1, t_2) \neq (0, 0, 0), \quad (t_1, t_2) \in [0, 2\pi] \times [0, 2\pi].$$

Thus, the considered torus is a regular surface; see Fig. 3.3, for the values $a = 3$, $b = 1/2$.

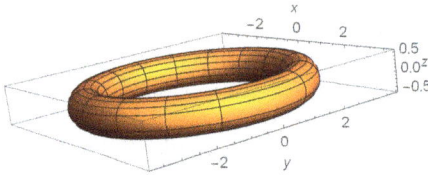

Figure 3.3: The torus $f(t_1, t_2) = ((3 + \frac{1}{2}\cos t_1)\cos t_2, (3 + \frac{1}{2}\cos t_1)\sin t_2, \frac{1}{2}\sin t_1), (t_1, t_2) \in [0, 2\pi] \times [0, 2\pi]$.

Exercise 3.1. Prove that the elliptic cylinder (see Fig. 3.4)

$$f(t_1, t_2) = (a\cos t_1, b\sin t_1, t_2), \quad t_1 \in [0, 2\pi], \quad t_2 \in [0, 2],$$

where $a, b \in \mathbb{R}$, $a, b > 0$, is a regular surface.

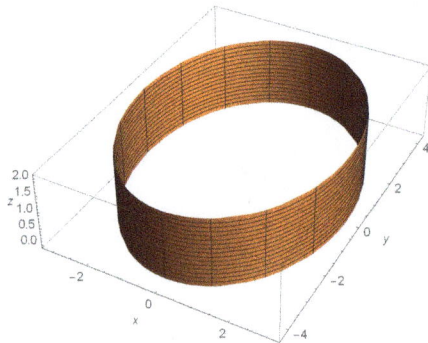

Figure 3.4: The elliptic cylinder $f(t_1, t_2) = (3\cos t_1, 4\sin t_1, t_2), t_1 \in [0, 2\pi], t_2 \in [0, 2]$.

Definition 3.3. The set $f(U) \subseteq \mathbb{R}^3$ is said to be the support of the parameterized surface (U, f).

Example 3.4. Let $U = \mathbb{R}^2$ and

$$f(t_1, t_2) = (t_2 + 2, 3t_1 + 4t_2 + 5, 6t_1 + 7t_2 + 8), \quad (t_1, t_2) \in \mathbb{R}^2$$

(see Fig. 3.5). We will find the support of f. Here

$$f_1(t_1, t_2) = t_2 + 2,$$

$$f_2(t_1, t_2) = 3t_1 + 4t_2 + 5,$$
$$f_3(t_1, t_2) = 6t_1 + 7t_2 + 8, \quad (t_1, t_2) \in \mathbb{R}^2.$$

Then

$$t_2 = f_1 - 2,$$
$$f_2 = 3t_1 + 4(f_1 - 2) + 5,$$
$$f_3 = 6t_1 + 7(f_1 - 2) + 8,$$

or

$$t_2 = f_1 - 2,$$
$$f_2 = 3t_1 + 4f_1 - 3,$$
$$f_3 = 6t_1 + 7f_1 - 6,$$

whereupon

$$t_2 = f_1 - 1,$$
$$t_1 = \frac{f_2 - 4f_1 + 3}{3},$$
$$t_1 = \frac{f_3 - 7f_1 + 6}{6}.$$

Hence,

$$\frac{f_2 - 4f_1 + 3}{3} = \frac{f_3 - 7f_1 + 6}{6}$$

and

$$2(f_2 - 4f_1 + 3) = f_3 - 7f_1 + 6,$$

or

$$2f_2 - 8f_1 - 6 = f_3 - 7f_1 - 6,$$

i. e., the support of the considered surface is the plane

$$f_1 - 2f_2 + f_3 = 0.$$

Example 3.5. Let $U = \mathbb{R}^2$ and

$$f(t_1, t_2) = (\cos t_1 \cos t_2, \cos t_1 \sin t_2, \sin t_1), \quad (t_1, t_2) \in \mathbb{R}^2.$$

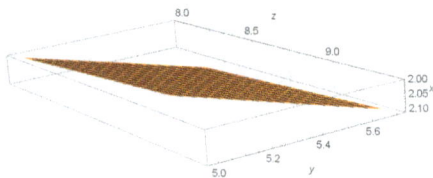

Figure 3.5: The line $f(t_1, t_2) = (t_2 + 2, 3t_1 + 4t_2 + 5, 6t_1 + 7t_2 + 8)$, $(t_1, t_2) \in (0, 1/10)$.

Here

$$f_1(t_1, t_2) = \cos t_1 \cos t_2,$$
$$f_2(t_1, t_2) = \cos t_1 \sin t_2,$$
$$f_3(t_1, t_2) = \sin t_1, \quad (t_1, t_2) \in \mathbb{R}^2.$$

Observe that

$$f_1^2 + f_2^2 + f_3^2 = (\cos t_1)^2 (\cos t_2)^2 + (\cos t_1)^2 (\sin t_2)^2 + (\sin t_1)^2$$
$$= (\cos t_1)^2 + (\sin t_1)^2$$
$$= 1,$$

i. e., the support of the considered surface is the sphere

$$f_1^2 + f_2^2 + f_3^2 = 1.$$

Exercise 3.2. Find the support of the surface

$$f(t_1, t_2) = (t_1 \cos t_2, t_1 \sin t_2, t_2), \quad (t_1, t_2) \in \mathbb{R}^2.$$

Answer 3.1. The helicoid

$$\tan f_3 = \frac{f_2}{f_1}.$$

Definition 3.4. Let $m, n \in \mathbb{N}$, $M \subseteq \mathbb{R}^m$, $N \subseteq \mathbb{R}^n$. A map $f : M \to N$ is called a homeomorphism if it is bijection, $f \in C(M)$, and its inverse $f^{-1} : N \to M$ is also continuous.

Definition 3.5. Let $m, n \in \mathbb{N}$, $M \subseteq \mathbb{R}^m$, $N \subseteq \mathbb{R}^n$. A differentiable map $f : M \to N$ is called a diffeomorphism if it is bijection and its inverse $f^{-1} : N \to M$ is differentiable. If $f \in C^r(M)$ and $f^{-1} \in C^r(N)$, then f is said to be C^r-diffeomorphism.

Example 3.6. Consider the map $f : \mathbb{R}^2 \to \mathbb{R}^2$ given by

$$f(t_1, t_2) = (t_1^2 + t_2^3, t_1^2 - t_2^3), \quad (t_1, t_2) \in \mathbb{R}^2, \quad t_1 t_2 \neq 0.$$

Here

$$f_1(t_1, t_2) = t_1^2 + t_2^3,$$
$$f_2(t_1, t_2) = t_1^2 - t_2^3, \quad (t_1, t_2) \in \mathbb{R}^2, \quad t_1 t_2 \neq 0.$$

Then

$$f_{1t_1}(t_1, t_2) = 2t_1,$$
$$f_{1t_2}(t_1, t_2) = 3t_2^2,$$
$$f_{2t_1}(t_1, t_2) = 2t_1,$$
$$f_{2t_2}(t_1, t_2) = -3t_2^2, \quad (t_1, t_2) \in \mathbb{R}^2, \quad t_1 t_2 \neq 0.$$

In the considered case, the Jacobian matrix is

$$J_f(t_1, t_2) = \begin{pmatrix} f_{1t_1}(t_1, t_2) & f_{1t_2}(t_1, t_2) \\ f_{2t_1}(t_1, t_2) & f_{2t_2}(t_1, t_2) \end{pmatrix}$$

$$= \begin{pmatrix} 2t_1 & 3t_2^2 \\ 2t_1 & -3t_2^2 \end{pmatrix}, \quad (t_1, t_2) \in \mathbb{R}^2, \quad t_1 t_2 \neq 0,$$

and for its determinant we have

$$\det J_f(t_1, t_2) = -6t_1 t_2^2 - 6t_1 t_2^2$$
$$= -12t_1 t_2^2, \quad (t_1, t_2) \in \mathbb{R}^2, \quad t_1 t_2 \neq 0.$$

From here, we conclude that the considered map is a diffeomorphism away from the t_1- and t_2-axis.

Example 3.7. Consider the map $f : \mathbb{R}^2 \to \mathbb{R}^2$ given by

$$f(t_1, t_2) = (\sin(t_1^2 + t_2^2), \cos(t_1^2 + t_2^2)), \quad (t_1, t_2) \in \mathbb{R}^2.$$

Here

$$f_1(t_1, t_2) = \sin(t_1^2 + t_2^2),$$
$$f_2(t_1, t_2) = \cos(t_1^2 + t_2^2), \quad (t_1, t_2) \in \mathbb{R}^2.$$

Then

$$f_{1t_1}(t_1, t_2) = 2t_1 \cos(t_1^2 + t_2^2),$$
$$f_{1t_2}(t_1, t_2) = 2t_2 \cos(t_1^2 + t_2^2),$$
$$f_{2t_1}(t_1, t_2) = -2t_1 \sin(t_1^2 + t_2^2),$$
$$f_{2t_2}(t_1, t_2) = -2t_2 \sin(t_1^2 + t_2^2), \quad (t_1, t_2) \in \mathbb{R}^2.$$

The Jacobian matrix is

$$J_f(t_1, t_2) = \begin{pmatrix} f_{1t_1}(t_1, t_2) & f_{1t_2}(t_1, t_2) \\ f_{2t_1}(t_1, t_2) & f_{2t_2}(t_1, t_2) \end{pmatrix}$$

$$= \begin{pmatrix} 2t_1 \cos(t_1^2 + t_2^2) & 2t_2 \cos(t_1^2 + t_2^2) \\ -2t_1 \sin(t_1^2 + t_2^2) & -2t_2 \sin(t_1^2 + t_2^2) \end{pmatrix}, \quad (t_1, t_2) \in \mathbb{R}^2,$$

and its determinant is

$$\det J_f(t_1, t_2) = -4t_1 t_2 \sin(t_1^2 + t_2^2) \cos(t_1^2 + t_2^2) + 4t_1 t_2 \sin(t_1^2 + t_2^2) \cos(t_1^2 + t_2^2)$$

$$= 0, \quad (t_1, t_2) \in \mathbb{R}^2.$$

Thus, the considered map is not a diffeomorphism.

Exercise 3.3. Prove that the map $f : [1, 2] \times [1, 2] \to \mathbb{R}^2$ given by

$$f(t_1, t_2) = (t_1 \cos t_2, t_1 \sin t_2), \quad (t_1, t_2) \in [1, 2] \times [1, 2],$$

is a diffeomorphism.

Definition 3.6. The two parameterized surfaces (U, f) and (V, g) are said to be equivalent if there is a diffeomorphism $\phi : U \to V$ such that

$$f = g(\phi).$$

Definition 3.7. A subset S of $\mathbb{R}^3$ is called a regular surface if for each point $a \in S$ there is a neighborhood W in S and a homeomorphism $f : U \to W$ so that $f(U) = W$ and (U, f) is a parameterized surface. The pair (U, f) is said to be a local parameterization of the surface S. The support $f(U)$ is called the domain of the parameterization. If $f(U) = S$, then the surface S is said to be a simple surface.

We have the following representations of the surfaces:
1. If (U, f) is a parametrized surface and

$$f(t_1, t_2) = (f_1(t_1, t_2), f_2(t_1, t_2), f_3(t_1, t_2)), \quad (t_1, t_2) \in U,$$

then the equations

$$f_1 = f_1(t_1, t_2),$$
$$f_2 = f_2(t_1, t_2),$$
$$f_3 = f_3(t_1, t_2), \quad (t_1, t_2) \in U,$$

are said to be parametric equations of the parameterized surface (U, f).
2. If $g : U \to \mathbb{R}, g \in C^1(U)$ and

$$f(t_1, t_2) = (t_1, t_2, g(t_1, t_2)),$$

then this representation of the parameterized surface (U, f) is said to be an explicit representation.

3. Let $W \subseteq \mathbb{R}^3$ and $F \in C^1(\mathbb{R}^3)$.

Definition 3.8. The set

$$S = \{(t_1, t_2, t_3) \in \mathbb{R}^3 : F(t_1, t_2, t_3) = 0\}$$

is said to be 0-level set of F.

Definition 3.9. The vector

$$\operatorname{grad} F(t_1, t_2, t_3) = (F_{t_1}(t_1, t_2, t_3), F_{t_2}(t_1, t_2, t_3), F_{t_3}(t_1, t_2, t_3)), \quad (t_1, t_2, t_3) \in W,$$

is said to be gradient vector of F.

Exercise 3.4. Let

$$\operatorname{grad} F(t_1, t_2, t_3) \neq (0, 0, 0), \quad (t_1, t_2, t_3) \in W.$$

Prove that S is a regular surface.

Solution. Let $(t_1^0, t_2^0, t_3^0) \in W$. Without loss of generality, suppose that

$$F_{t_3}(t_1^0, t_2^0, t_3^0) \neq 0.$$

Then, by the implicit function theorem, there is a neighborhood M of (t_1^0, t_2^0, t_3^0) and $f \in C^1$ such that

$$t_3 = f(t_1, t_2), \quad (t_1, t_2, t_3) \in M.$$

This completes the proof.

We call

$$F(t_1, t_2, t_3) = 0$$

the implicit representation of the parameterized surface

$$t_3 = f(t_1, t_2), \quad (t_1, t_2, t_3) \in M.$$

Example 3.8. We will find the parametric equations of the ellipsoid

$$\frac{x^2}{a^2} + \frac{y^2}{b^2} + \frac{z^2}{c^2} = 1,$$

where $a, b, c \in \mathbb{R}$, $(a, b, c) \neq (0, 0, 0)$. Let

$$x = a \cos t_1 \cos t_2,$$
$$y = b \sin t_1 \cos t_2,$$
$$z = c \sin t_2, \quad t_1 \in [0, 2\pi], \quad t_2 \in [0, \pi].$$

We have

$$
\begin{aligned}
\frac{x^2}{a^2} + \frac{y^2}{b^2} + \frac{z^2}{c^2} &= \frac{1}{a^2}(a \cos t_1 \cos t_2)^2 + \frac{1}{b^2}(b \sin t_1 \cos t_2)^2 + \frac{1}{c^2}(c \sin t_2)^2 \\
&= (\cos t_1)^2(\cos t_2)^2 + (\sin t_1)^2(\cos t_2)^2 + (\sin t_2)^2 \\
&= ((\cos t_1)^2 + (\sin t_1)^2)(\cos t_2)^2 + (\sin t_2)^2 \\
&= (\cos t_2)^2 + (\sin t_2)^2 \\
&= 1, \quad t_1 \in [0, 2\pi], \quad v \in [0, \pi].
\end{aligned}
$$

Thus, the parametric equation of the ellipsoid is

$$f(t_1, t_2) = (a \cos t_1 \cos t_2, b \sin t_1 \cos t_2, c \sin t_2), \quad t_1 \in [0, 2\pi], \quad t_2 \in [0, \pi].$$

See Fig. 3.6 for the values $a = 1$, $b = 2$, and $c = 3$.

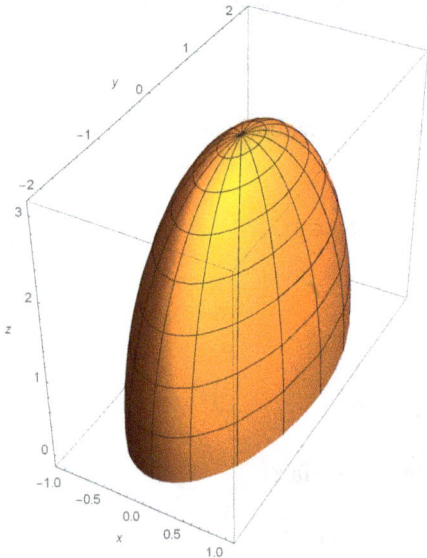

Figure 3.6: The ellipsoid $f(t_1, t_2) = (\cos t_1 \cos t_2, 2 \sin t_1 \cos t_2, 3 \sin t_2)$, $t_1 \in [0, 2\pi]$, $t_2 \in [0, \pi]$.

Example 3.9. Consider the hyperboloid of one sheet

$$\frac{x^2}{a^2} + \frac{y^2}{b^2} - \frac{z^2}{c^2} = 1,$$

where $a, b, c \in \mathbb{R}$, $(a, b, c) \neq (0, 0, 0)$. Let

$$x = a \cosh t_1 \cos t_2,$$
$$y = b \cosh t_1 \sin t_2,$$
$$z = c \sinh t_1, \quad t_1 \in \mathbb{R}, \quad t_2 \in [0, 2\pi].$$

Then

$$
\begin{aligned}
\frac{x^2}{a^2} + \frac{y^2}{b^2} - \frac{z^2}{c^2} &= \frac{1}{a^2}(a \cosh t_1 \cos t_2)^2 + \frac{1}{b^2}(b \cosh t_1 \sin t_2)^2 - \frac{1}{c^2}(c \sinh t_1)^2 \\
&= (\cosh t_1)^2(\cos t_2)^2 + (\cosh t_1)^2(\sin t_2)^2 - (\sinh t_1)^2 \\
&= (\cosh t_1)^2((\cos t_2)^2 + (\sin t_2)^2) - (\sinh t_1)^2 \\
&= (\cosh t_1)^2 - (\sinh t_1)^2 \\
&= 1.
\end{aligned}
$$

Thus, the parametric equation of the hyperboloid of one sheet is

$$f(t_1, t_2) = (a \cosh t_1 \cos t_2, b \cosh t_1 \sin t_2, c \sinh t_1), \quad t_1 \in \mathbb{R}, \quad t_2 \in [0, 2\pi].$$

See Fig. 3.7 for the values $a = 1$, $b = 2$, and $c = 3$.

Example 3.10. Consider the two-sheeted hyperboloid

$$\frac{x^2}{a^2} + \frac{y^2}{b^2} - \frac{z^2}{c^2} = -1,$$

where $a, b, c \in \mathbb{R}$, $(a, b, c) \neq (0, 0, 0)$. Let

$$x = a \sinh t_1 \cos t_2,$$
$$y = b \sinh t_1 \sin t_2,$$
$$z = c \cosh t_1, \quad t_1 \in \mathbb{R}, \quad t_2 \in [0, 2\pi].$$

Then

$$
\begin{aligned}
\frac{x^2}{a^2} + \frac{y^2}{b^2} - \frac{z^2}{c^2} &= \frac{1}{a^2}(a \sinh t_1 \cos t_2)^2 + \frac{1}{b^2}(\sinh t_1 \sin t_2)^2 - \frac{1}{c^2}(c \cosh t_1)^2 \\
&= (\sinh t_1)^2(\cos t_2)^2 + (\sinh t_1)^2(\sin t_2)^2 - (\cosh t_1)^2 \\
&= (\sinh t_1)^2((\cos t_2)^2 + (\sin t_2)^2) - (\cosh t_1)^2
\end{aligned}
$$

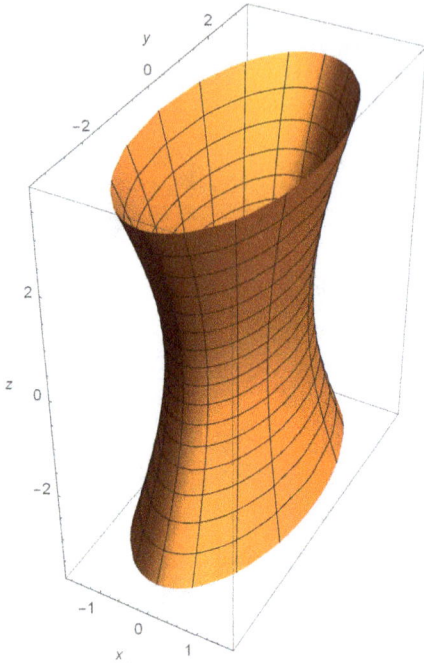

Figure 3.7: The hyperboloid of one sheet $f(t_1, t_2) = (\cosh t_1 \cos t_2, 2\cosh t_1 \sin t_2, 3\sinh t_1)$, $t_1 \in [-1,1]$, $t_2 \in [0, 2\pi]$.

$$= (\sinh t_1)^2 - (\cosh t_1)^2$$
$$= -1.$$

Thus, the parametric equation of the two-sheeted hyperboloid is

$$f(t_1, t_2) = (a\sinh t_1 \cos t_2, b\sinh t_1 \sin t_2, c\cosh t_1), \quad t_1 \in \mathbb{R}, \quad t_2 \in [0, 2\pi].$$

See Fig. 3.8 for the values $a = 1$, $b = 2$, and $c = 3$.

Example 3.11. Consider the elliptic paraboloid

$$\frac{x^2}{a^2} + \frac{y^2}{b^2} = z,$$

where $a, b \in \mathbb{R}$, $(a, b) \neq (0, 0)$. Let

$$x = at_1 \cos t_2,$$
$$y = bt_1 \sin t_2,$$
$$z = t_1^2, \quad t_1 \in \mathbb{R}, \quad t_2 \in [0, 2\pi].$$

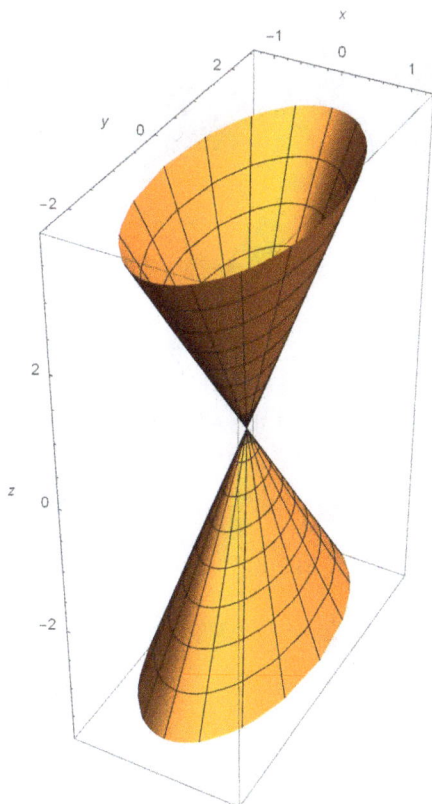

Figure 3.8: The two-sheeted hyperboloid $f(t_1, t_2) = (\cosh t_1 \cos t_2, 2 \cosh t_1 \sin t_2, 3 \sinh t_1)$, $t_1 \in [-1, 1]$, $t_2 \in [0, 2\pi]$.

Then

$$
\begin{aligned}
\frac{x^2}{a^2} + \frac{y^2}{b^2} &= \frac{1}{a^2}(at_1 \cos t_2)^2 + \frac{1}{b^2}(bt_1 \sin t_2)^2 \\
&= t_1^2(\cos t_2)^2 + t_1^2(\sin t_2)^2 \\
&= t_1^2((\cos t_2)^2 + (\sin t_2)^2) \\
&= t_1^2 \\
&= z, \quad t_1 \in \mathbb{R}, \quad t_2 \in [0, 2\pi].
\end{aligned}
$$

Thus, the parametric equation of the elliptic paraboloid is

$$
f(t_1, t_2) = (at_1 \cos t_2, bt_1 \sin t_2, t_1^2), \quad t_1 \in \mathbb{R}, \quad t_2 \in [0, 2\pi].
$$

See Fig. 3.9 for the values $a = 1$ and $b = 2$.

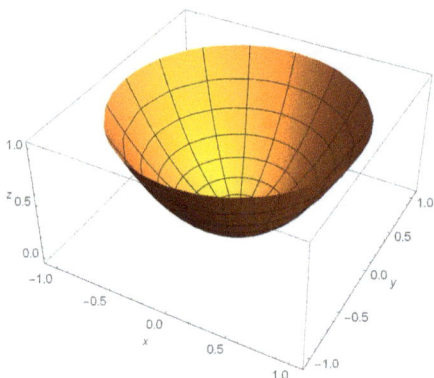

Figure 3.9: The elliptic paraboloid $f(t_1, t_2) = (t_1 \cos t_2, 2t_1 \sin t_2, t_1^2)$, $t_1 \in [-1, 1]$, $t_2 \in [0, 2\pi]$.

Example 3.12. Consider the elliptic cylinder

$$\frac{x^2}{a^2} + \frac{y^2}{b^2} = 1,$$

where $a, b \in \mathbb{R}$, $(a, b) \neq (0, 0)$. Let

$$x = a \cos t_1,$$
$$y = a \sin t_1,$$
$$z = v, \quad t_1 \in [0, 2\pi], \quad t_2 \in \mathbb{R}.$$

Then

$$\frac{x^2}{a^2} + \frac{y^2}{b^2} = \frac{1}{a^2}(a \cos t_1)^2 + \frac{1}{b^2}(b \sin t_1)^2$$
$$= (\cos t_1)^2 + (\sin t_1)^2$$
$$= 1, \quad t_1 \in [0, 2\pi], \quad t_2 \in \mathbb{R}.$$

Thus, the parametric equation of the elliptic cylinder is

$$f(t_1, t_1) = (a \cos t_1, b \sin t_1, t_2), \quad t_1 \in [0, 2\pi], \quad t_2 \in \mathbb{R}.$$

See Fig. 3.10 for the values $a = 1$ and $b = 2$.

Example 3.13. Consider the cone

$$\frac{x^2}{a^2} + \frac{y^2}{b^2} = \frac{z^2}{c^2},$$

where $a, b, c \in \mathbb{R}$, $(a, b, c) \neq (0, 0, 0)$. Let

$$x = at_1 \cos t_2,$$

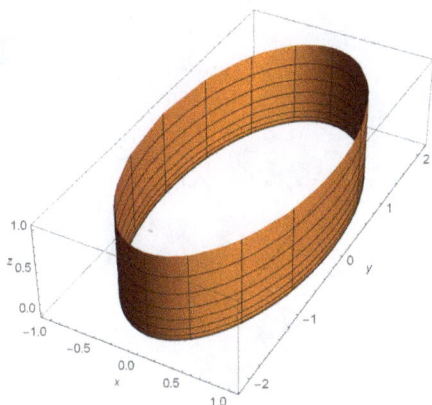

Figure 3.10: The elliptic cylinder $f(t_1, t_2) = (a \cos t_1, 2 \sin t_1, t_2), t_1 \in [-1, 1], t_2 \in [0, 2\pi]$.

$$y = bt_1 \sin t_2,$$
$$z = ct_1, \quad t_1 \in \mathbb{R}, \quad t_2 \in [0, 2\pi].$$

Then

$$\frac{x^2}{a^2} + \frac{y^2}{b^2} = \frac{1}{a^2}(at_1 \cos v)^2 + \frac{1}{b^2}(bt_1 \sin t_2)^2$$
$$= t_1^2(\cos t_2)^2 + t_1^2(\sin t_2)^2$$
$$= t_1^2((\cos t_2)^2 + (\sin t_2)^2)$$
$$= t_1^2$$
$$= \frac{z^2}{c^2}.$$

Thus, the parametric equation of the cone is

$$f(t_1, t_2) = (at_1 \cos t_2, bt_1 \sin t_2, ct_1), \quad u \in \mathbb{R}, \quad t_2 \in [0, 2\pi].$$

See Fig. 3.11 for the values $a = 1$, $b = 1$, and $c = 2$.

Example 3.14. Consider the hyperbolic paraboloid

$$\frac{x^2}{a^2} - \frac{y^2}{b^2} = 2z,$$

where $a, b \in \mathbb{R}, (a, b) \neq (0, 0)$. Let

$$x = a(t_1 + t_2),$$
$$y = b(t_2 - t_1),$$
$$z = 2t_1 t_2, \quad (t_1, t_2) \in \mathbb{R}^2.$$

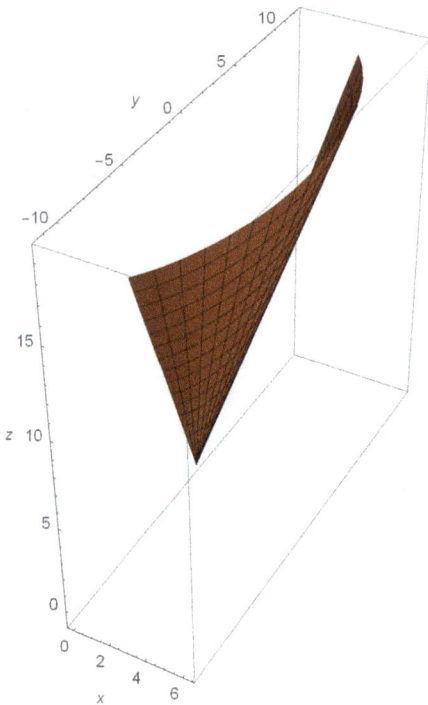

Figure 3.11: The cone $f(t_1, t_2) = (t_1 \cos t_2, 2t_1 \sin t_2, 3t_1)$, $t_1 \in [-1, 1]$, $t_2 \in [0, 2\pi]$.

We have

$$\frac{x^2}{a^2} - \frac{y^2}{b^2} = \frac{1}{a^2}(a(t_1 + t_2))^2 - \frac{1}{b^2}(b(t_2 - t_1))^2$$
$$= (t_1 + t_2)^2 - (t_2 - t_1)^2$$
$$= t_1^2 + 2t_1t_2 + t_2^2 - (t_2^2 - 2t_1t_2 + t_1^2)$$
$$= 4t_1t_2$$
$$= 2z, \quad (u, v) \in \mathbb{R}^2.$$

Thus, the parametric equation of the hyperbolic paraboloid is

$$f(t_1, t_2) = (a(t_1 + t_2), b(t_2 - t_1), 2t_1t_2), \quad (t_1, t_2) \in \mathbb{R}^2.$$

See Fig. 3.12 for the values $a = 1$ and $b = 2$.

Exercise 3.5. Prove that the equations

$$f(u, v) = \left(\frac{u}{u^2 + v^2}, \frac{v}{u^2 + v^2}, \frac{1}{u^2 + v^2} \right), \quad (u, v) \in \mathbb{R}^2,$$

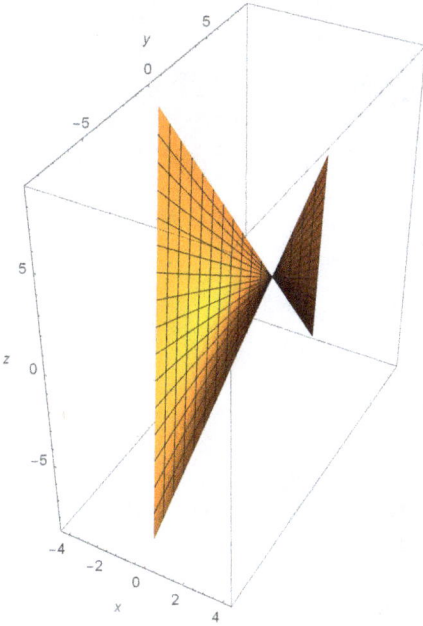

Figure 3.12: The cone $f(t_1, t_2) = (t_1 + t_2, 2(t_2 - t_1), 2t_1 t_2), t_1 \in [-1, 1], t_2 \in [-1, 1]$.

and

$$g(t_1, t_2) = (t_1 \cos t_2, t_1 \sin t_2, t_1^2), \quad t_1 \in \mathbb{R}, \quad t_2 \in [0, 2\pi],$$

determine the same surface.

3.2 The equivalence of local representations

Definition 3.10. Let S be a surface, (U, f) be its local parameterization, and $W = f(U)$. Then the map $f^{-1} : W \to U$ is a bijection and is called a curvilinear coordinate system on S, or a chart on S.

Exercise 3.6. Let (U, f) be a local parameterization of the surface S, $f(U) = W$, and $f^{-1} : W \to U$. Then for each point $a \in W$, there is an open set B in the topology of $\mathbb{R}^3$ and a smooth map $G : B \to V$ such that $a \in B$, $V \subset U$ is an open subset, and

$$f^{-1}_{|_{W \cap B}} = G_{|_{W \cap B}}.$$

Solution. Let

$$f(t_1, t_2) = (f_1(t_1, t_2), f_2(t_1, t_2), f_3(t_1, t_2)), \quad (t_1, t_2) \in U,$$

and

$$a = f(t_1^0, t_2^0).$$

Note that

$$\text{rank} \begin{pmatrix} f_{1t_1} & f_{1t_2} \\ f_{2t_1} & f_{2t_2} \\ f_{3t_1} & f_{3t_2} \end{pmatrix} = 2.$$

Without loss of generality, suppose that

$$\begin{vmatrix} f_{1t_1} & f_{1t_2} \\ f_{2t_1} & f_{2t_2} \end{vmatrix} \neq 0.$$

Then, by the inverse function theorem, it follows that there is an open neighborhood V of the point (t_1^0, t_2^0) in U and an open neighborhood V_1 of the point $(f_1(t_1^0, t_2^0), f_2(t_1^0, t_2^0))$ such that $f : V \to V_1$ is a diffeomorphism. Because $f : U \to W$ is a homeomorphism, we have that $f(V)$ is an open neighborhood in S of the point a. Therefore there is an open neighborhood of the point a such that

$$f(V) = B \cap S = B \cap W.$$

Now, define the map $\phi : \mathbb{R}^3 \to \mathbb{R}^2$ as follows:

$$\phi(t_1, t_2, t_3) = (t_1, t_2), \quad (t_1, t_2) \in \mathbb{R}^2.$$

Let

$$G = (f^{-1}(\phi))\big|_B : B \to U.$$

Note that G is a smooth map. Next, to each point $(s_1, s_2, s_3) \in B \cap W$ corresponds a single point

$$(t_1, t_2) = f^{-1}(s_1, s_2, s_3) \in V.$$

Also, to each point $(s_1, s_2) \in V_1$ corresponds the point

$$(t_1, t_2) = f^{-1}(s_1, s_2) \in V.$$

Thus, if $(s_1, s_2, s_3) \in B \cap W$, then

$$f^{-1}(s_1, s_2, s_3) = (t_1, t_2)$$
$$= f^{-1}(s_1, s_2)$$

$$= f^{-1}(\phi(s_1, s_2, s_3))$$
$$= G(s_1, s_2, s_3).$$

This completes the solution.

Exercise 3.7. Let (U, f) and (U_1, f_1) be two local parameterizations of a surface S and $f(U) = f_1(U_1)$. Then there is a diffeomorphism $\phi : U \to U_1$ such that

$$f = f_1(\phi).$$

Solution. Let $W = f(U) = f_1(U_1)$ and

$$\phi = f_1^{-1} \circ f.$$

Then $\phi : U \to U_1$. Since $f_1 : U_1 \to W$ is a homeomorphism, we have that $f_1^{-1} : W \to U_1$ is a homeomorphism. Therefore $\phi : U \to U_1$ is a homeomorphism. Now, we will prove that each point $(t_1^0, t_2^0) \in U$ has an open neighborhood $V \subset U$ such that the map $\phi|_V$ is smooth. Let

$$a = f_1(\phi(t_1^0, t_2^0)).$$

By Exercise 3.6, it follows that there is an open set B of $\mathbb{R}^3$ such that $G : B \to U_1$ is a smooth map so that

$$f_1^{-1}\Big|_{B \cap W} = G\Big|_{B \cap W}$$

and

$$V = f^{-1}(B \cap W).$$

Then

$$\phi|_V = f_1^{-1} \cdot f|_V = G \circ f|_V$$

and $\phi|_V$ is a smooth map. As above, ϕ^{-1} is a smooth map. This completes the solution.

Exercise 3.8. Let (U, f) be a regular parametrized surface. Then each point $(t_1^0, t_2^0) \in U$ has an open neighborhood $V \subset U$ such that $f(V)$ is a simple surface in $\mathbb{R}^3$ for which $(V, f|_V)$ is a global representation.

Solution. Let

$$f(t_1, t_2) = (f_1(t_1, t_2), f_2(t_1, t_2), f_3(t_1, t_2)), \quad (t_1, t_2) \in U,$$

and

$$\begin{vmatrix} f_{1t_1}(t_1^0, t_2^0) & f_{1t_2}(t_1^0, t_2^0) \\ f_{2t_1}(t_1^0, t_2^0) & f_{2t_2}(t_1^0, t_2^0) \end{vmatrix} \neq 0.$$

Then, by the inverse function theorem, there is an open neighborhood $V \subset U$ of the point (t_1^0, t_2^0) and an open neighborhood V_1 of the point $(s_{10}, s_{20}, s_{30}) = f(t_1^0, t_2^0)$ such that $f : V \to V_1$ is a diffeomorphism. Let now $(t_1, t_2), (y_1, y_2) \in V$ be such that

$$f_1(t_1, t_2) = f_1(y_1, y_2),$$
$$f_2(t_1, t_2) = f_2(y_1, y_2).$$

Since $f : V \to V_1$ is a diffeomorphism, it is an injective map and then

$$(t_1, t_2) = (y_1, y_2).$$

Since $f : U \to \mathbb{R}^3$ is continuous, we have that $f_{|_V} : V \to f(V)$ is continuous. Note that

$$(x, y, z) \in f(V) \to (u, v, w) \in V_1 \to (t_1, t_2) = f^{-1}(u, v, w).$$

Thus, the inverse map of $f_{|_V} : V \to f(V)$ is continuous. This completes the solution.

3.3 Curves on surfaces

Exercise 3.9. Let (U, f) be a parameterization of the surface S and $(I, g = g(t))$ be a smooth parameterized curve whose support is included in $f(U)$. Then there is a unique smooth parameterized curve (I, g_1) on U such that

$$g(t) = f(g_1(t)), \quad t \in I. \tag{3.2}$$

Conversely, any smooth parametrized curve g_1 on U defines, through (3.2), a smooth curve on $f(U)$. The regularity of g at t is equivalent to the regularity of g_1 at t.

Solution. Since $f : U \to f(U)$ is a homeomorphism and $g(I) \subset f(U)$, we get

$$g_1 = f^{-1} \circ g.$$

We have that g_1 is continuous because it is a composition of two continuous maps. Let $t \in I$. Then $g(t) \in f(U)$. By Exercise 3.6, it follows that there is an open neighborhood B of the point $g(t)$ and a smooth map $G : B \to U$ such that

$$f^{-1}_{|_{B \cap f(U)}} = G_{|_{B \cap f(U)}}.$$

Therefore, the map g_1 can be represented in a neighborhood of the point t as a composition of g and G. Since G and g are smooth, we have that g_1 is smooth. For the converse assertion, by (3.2) and the smoothness of f and g_1, it follows that g is smooth. Let

$$g_1(t) = (u(t), v(t)), \quad t \in I.$$

Then, by (3.2), we find

$$g(t) = f(u(t), v(t)), \quad t \in I.$$

Now, we differentiate the latter equation with respect to t and find

$$g'(t) = f_{t_1}(u(t), v(t))u'(t) + f_{t_2}(u(t), v(t))v'(t), \quad t \in I.$$

Since

$$f_{t_1} \times f_{t_2} \neq 0 \quad \text{on } U,$$

we have that f_{t_1} and f_{t_2} are not collinear. Therefore

$$g'(t) = 0, \quad t \in I,$$

if and only if

$$g_1'(t) = 0, \quad t \in I.$$

This completes the solution.

Definition 3.11. Let U, f, g, g_1, and I be as in Exercise 3.9. Then the parameterized curve g_1 on U is called a local parameterization of g in (U, f). The equations

$$u = u(t),$$
$$v = v(t), \quad t \in I,$$

are called local equations of g.

3.4 The tangent vector space, tangent plane, and normal to a surface

Let $a \in \mathbb{R}^3$. Denote by $\mathbb{R}_a^3$ the space of all vectors with the origin a.

Definition 3.12. A vector $w \in \mathbb{R}^3$ is called a tangent vector to the surface S if there is a parameterized curve $(I, g = g(t))$ on S and $t_0 \in I$ such that $g(t_0) = a$ and

$$w = g'(t_0).$$

Thus, a tangent vector to a surface is a tangent vector to a parameterized curve on the surface S.

By T_aS we will denote the set of all tangent vectors to S at a. If

$$g(t) = f(u(t), v(t)), \quad t \in I,$$

then

$$g'(t) = f_u(u(t), v(t))u'(t) + f_v(u(t), v(t))v'(t), \quad t \in I.$$

Exercise 3.10. The set T_aS is a two-dimensional vector subspace of $\mathbb{R}^3$. If (U, f) is a local parameterization of S, $a = f(u_0, v_0)$, then $f_u(u_0, v_0)$ and $f_v(u_0, v_0)$ become a basis to T_aS.

Solution. Let $(I, g = g(t))$ be a parameterized curve on S and $g(t_0) = a$ for some $t_0 \in I$. Assume that $g(I) \subset f(U)$ and

$$g(t) = f(u(t), v(t)), \quad t \in I.$$

Then

$$g'(t) = f_u(u(t), v(t))u'(t) + f_v(u(t), v(t))v'(t), \quad t \in I.$$

Note that any vector in the form

$$w = \alpha f_u(u_0, v_0) + \beta f_v(u_0, v_0),$$

for some $\alpha, \beta \in \mathbb{R}$, is a tangent vector to the curve with equations

$$u = u_0 + \alpha t,$$
$$v = v_0 + \beta t,$$

which is a curve on S passing through the point a for $t = t_0$. Thus, $w \in T_aS$. This completes the solution.

Definition 3.13. The vector space T_aS is called the tangent space to S at a. The plane passing through a and having T_aS as a directing plane is called the tangent plane of S at a.

If

$$a = (f_1(t_1^0, t_2^0), f_2(t_1^0, t_2^0), f_3(t_1^0, t_2^0)),$$

then the equation of the tangent plane is given by

$$\begin{vmatrix} X - f_1(t_1^0, t_2^0) & Y - f_2(t_1^0, t_2^0) & Z - f_3(t_1^0, t_2^0) \\ f_{1t_1}(t_1^0, t_2^0) & f_{2t_1}(t_1^0, t_2^0) & f_{3t_1}(t_1^0, t_2^0) \\ f_{1t_2}(t_1^0, t_2^0) & f_{2t_2}(t_1^0, t_2^0) & f_{3t_2}(t_1^0, t_2^0) \end{vmatrix} = 0$$

or, equivalently,

$$(X - f_1(t_1^0, t_2^0), Y - f_2(t_1^0, t_2^0), Z - f_3(t_1^0, t_2^0)) \cdot (f_{t_1}(t_1^0, t_2^0) \times f_{t_2}(t_1^0, t_2^0)) = 0.$$

If the surface S is given by

$$z = f(x, y), \quad (x, y) \in U,$$

where $U \subset \mathbb{R}^2$ and $f \in C^1(U)$, then the equation of the tangent plane at (x_0, y_0, z_0), $(x_0, y_0) \in U, z_0 = f(x_0, y_0)$, is given by

$$z - z_0 = f_x(x_0, y_0)(x - x_0) + f_y(x_0, y_0)(y - y_0).$$

If the surface S is given by

$$F(x, y, z) = 0, \quad (x, y, z) \in V,$$

where $V \subset \mathbb{R}^3, F \in C^1(V)$, then the equation of the tangent plane at $(x_0, y_0, z_0) \in V$ is

$$F_x(x_0, y_0, z_0)(x - x_0) + F_y(x_0, y_0, z_0)(y - y_0) + F_z(x_0, y_0, z_0)(z - z_0) = 0.$$

Example 3.15. We will find the equation of the tangent plane to the surface

$$z = xy, \quad (x, y) \in \mathbb{R}^2,$$

at the point $(2, 1, 2)$. We have

$$z_x(x, y) = y,$$
$$z_y(x, y) = x, \quad (x, y) \in \mathbb{R}^2,$$
$$z_x(2, 1) = 1,$$
$$z_y(2, 1) = 2.$$

Then the equation of the tangent plane is

$$z - 2 = (x - 2) + 2(y - 1)$$

or

$$x + 2y - z = 2;$$

see Fig. 3.13.

Example 3.16. We will find the equation of the tangent plane to the surface

$$x^2 + y^2 + z^2 = 169, \quad (x, y, z) \in \mathbb{R}^3,$$

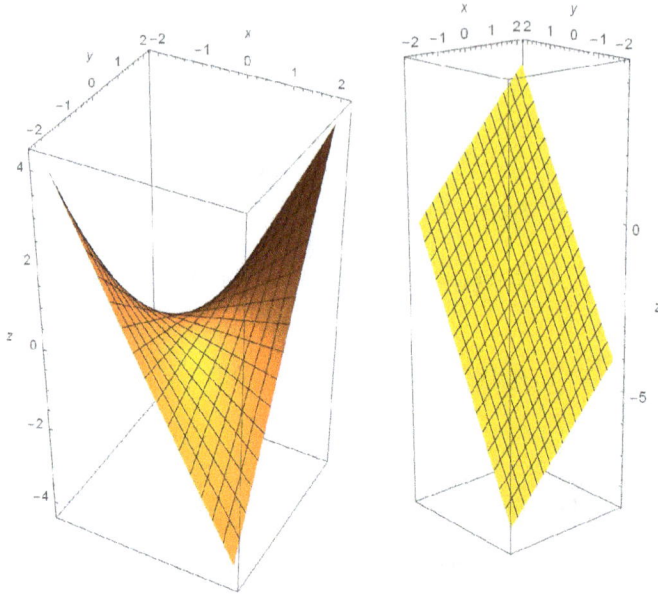

Figure 3.13: The surface $z = xy$ (left) and its tangent plane $x + 2y - z = 2$ (right) at $(2, 1, 2)$, $x, y \in [-2, 2]$.

at the point $(3, 4, -12)$. Let

$$F(x, y, z) = x^2 + y^2 + z^2 - 169, \quad (x, y, z) \in \mathbb{R}^3.$$

Then

$$F_x(x, y, z) = 2x,$$
$$F_y(x, y, z) = 2y,$$
$$F_z(x, y, z) = 2z, \quad (x, y, z) \in \mathbb{R}^3,$$

and

$$F_x(3, 4, -12) = 6,$$
$$F_y(3, 4, -12) = 8,$$
$$F_z(3, 4, -12) = -24.$$

Thus, the equation of the tangent plane is

$$6(x - 3) + 8(y - 4) - 24(z + 12) = 0$$

or

$$3x + 4y - 12z = 169;$$

see Fig. 3.14.

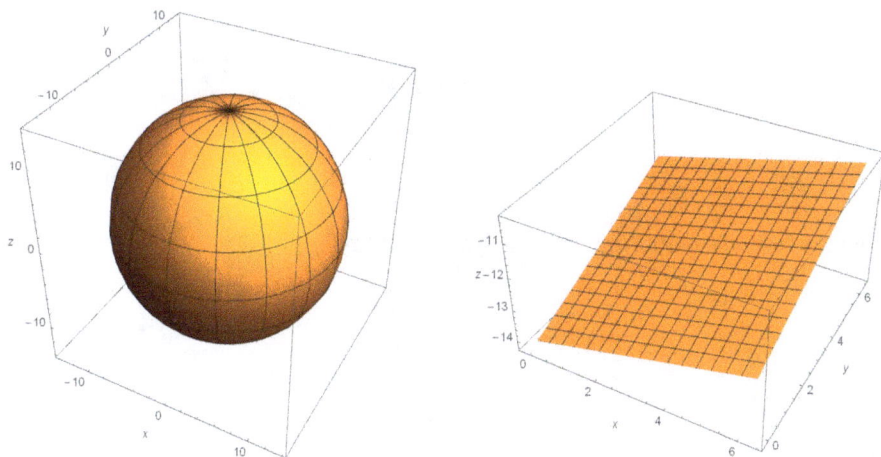

Figure 3.14: The sphere $x^2 + y^2 + z^2 = 169$ (left) and its tangent plane $3x + 4y - 12z = 169$ (right) at $(3, 4, -12)$.

Example 3.17. We will find the equation of the tangent plane to the surface

$$f(t_1, t_2) = (t_1 + t_2, t_1^2 + t_2^2, t_1^3 + t_2^3), \quad (t_1, t_2) \in \mathbb{R}^2,$$

at the point $(3, 5, 9)$. Here

$$f_1(t_1, t_2) = t_1 + t_2,$$
$$f_2(t_1, t_2) = t_1^2 + t_2^2,$$
$$f_3(t_1, t_2) = t_1^3 + t_2^3, \quad (t_1, t_2) \in \mathbb{R}^2.$$

Then

$$f_{1t_1}(t_1, t_2) = 1,$$
$$f_{1t_2}(t_1, t_2) = 1,$$
$$f_{2t_1}(t_1, t_2) = 2t_1,$$
$$f_{2t_2}(t_1, t_2) = 2t_2,$$
$$f_{3t_1}(t_1, t_2) = 3t_1^2,$$
$$f_{3t_2}(t_1, t_2) = 3t_2^2, \quad (t_1, t_2) \in \mathbb{R}^2.$$

Next, we have

$$t_1 + t_2 = 3,$$
$$t_1^2 + t_2^2 = 5,$$
$$t_1^3 + t_2^3 = 9$$

or

$$t_2 = 3 - t_1,$$
$$t_1^2 + (3 - t_1)^2 = 5,$$
$$t_1^3 + (3 - t_1)^3 = 9.$$

From the second equation of the latter system, we find

$$5 = t_1^2 + (3 - t_1)^2$$
$$= t_1^2 + 9 - 6t_1 + t_1^2$$
$$= 2t_1^2 - 6t_1 + 9,$$

whereupon

$$2t_1^2 - 6t_1 + 4 = 0,$$

or

$$t_1^2 - 3t_1 + 2 = 0.$$

Thus,

$$(t_1^1, t_2^1) = (1, 2), \quad (t_1^2, t_2^2) = (2, 1).$$

From here, we get

$$f_{1t_1}(t_1^1, t_2^1) = 1,$$
$$f_{1t_2}(t_1^1, t_2^1) = 1,$$
$$f_{2t_1}(t_1^1, t_2^1) = 2,$$
$$f_{2t_2}(t_1^1, t_2^1) = 4,$$
$$f_{3t_1}(t_1^1, t_2^1) = 3,$$
$$f_{3t_2}(t_1^1, t_2^1) = 12$$

and

$$f_{1t_1}(t_1^2, t_2^2) = 1,$$
$$f_{1t_2}(t_1^2, t_2^2) = 1,$$

$$f_{2t_1}(t_1^2, t_2^2) = 4,$$
$$f_{2t_2}(t_1^2, t_2^2) = 2,$$
$$f_{3t_1}(t_1^2, t_2^2) = 12,$$
$$f_{3t_2}(t_1^2, t_2^2) = 3.$$

The equation of the tangent plane is

$$0 = \begin{vmatrix} x-3 & y-5 & z-9 \\ 1 & 2 & 3 \\ 1 & 4 & 12 \end{vmatrix}$$

$$= 24(x-3) + 3(y-5) + 4(z-9) - 2(z-9) - 12(x-3) - 12(y-5),$$

or

$$12x - 9y + 2z = 9;$$

see Fig. 3.15.

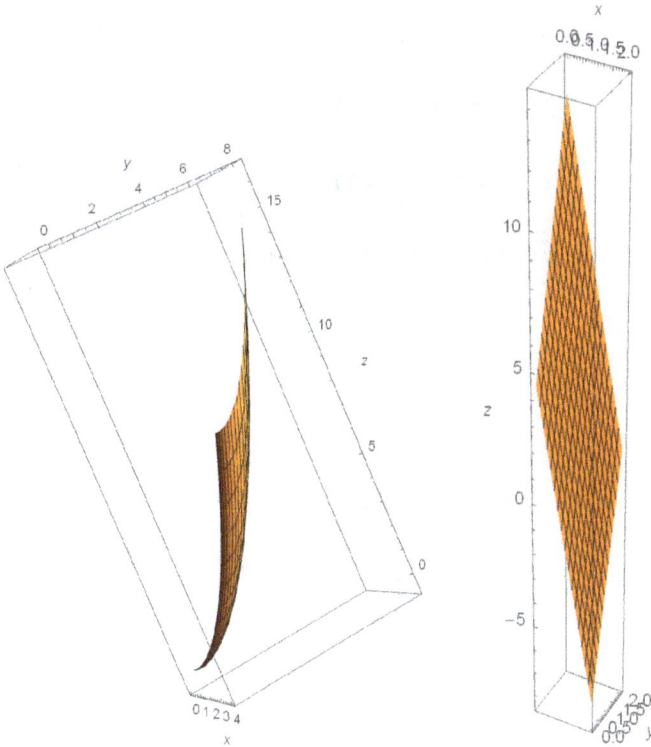

Figure 3.15: The surface $f(t_1, t_2) = (t_1 + t_2, t_1^2 + t_2^2, t_1^3 + t_2^3)$ (left) and its tangent plane $12x - 9y + 2z = 9$ (right) at $(3, 5, 9)$, $t_1, t_2 \in [0, 2]$.

Exercise 3.11. Find the equation of the tangent plane to the following surfaces at the given points:

1.
$$z = x^2 + y^2, \quad (1,1,2);$$

2.
$$z = 2x^2 - 4y^2, \quad (-2,1,4);$$

3.
$$z = (x-y)^2 - x + 2y, \quad (1,1,1);$$

4.
$$z = x^3 - 3xy + y^3, \quad (1,1,-1);$$

5.
$$z = \sqrt{x^2 + y^2} - xy, \quad (-3,4,17);$$

6.
$$xy^2 + z^3 = 12, \quad (1,2,2);$$

7.
$$x^3 + y^3 + z^3 + xyz = 6, \quad (1,2,-1);$$

8.
$$xyz(z^2 - x^2) = 6 + y^5, \quad (1,2,-1);$$

9.
$$\sqrt{x^2 + y^2 + z^2} = x + y + z - 4, \quad (2,3,6);$$

10.
$$e^z - z + xy = 3, \quad (2,1,0);$$

11.
$$(t_1, t_1^2 - 2t_1t_2, t_1^3 - 3t_1^2t_2), \quad (1,3,4).$$

Answer 3.2. 1.
$$2x + 2y - z = 2;$$

2.
$$8x + 8y + z + 4 = 0;$$

3.
$$x - 2y + z = 0;$$

4.
$$z = -1;$$

5.
$$23x - 19y + 5z = -60;$$

6.
$$x + y + 3z = 9;$$

7.
$$x + 11y + 5z = 18;$$

8.
$$x - 20y + z = -40;$$

9.
$$5x + 4y + z = 28;$$

10.
$$x + 2y = 4;$$

11.
$$6x + 3y - 2z = -11.$$

Let now $(U, f = f(t_1, t_2))$ be a parameterized surface and $(t_1^0, t_2^0) \in U$. Then

$$f(t_1^0 + ah, t_2^0 + \beta h) = f(t_1^0, t_2^0) + h(af_{t_1}(t_1^0, t_2^0) + \beta f_{t_2}(t_1^0, t_2^0)) + h\varepsilon,$$

where $a, \beta \in \mathbb{R}$ and

$$\lim_{h \to 0} \varepsilon = 0.$$

Let Π be a plane in $\mathbb{R}^3$ passing through the point $p_0 = f(t_1^0, t_2^0)$ and d be the distance from the point

$$p = f(t_1^0 + ah, t_2^0 + \beta h)$$

to the plane Π and δ be the distance between p_0 and p.

Exercise 3.12. Prove that the plane Π is the tangent plane to S at p_0 if and only if for any $a, \beta \in \mathbb{R}, a^2 + \beta^2 \neq 0$, we have

$$\lim_{h \to 0} \frac{d}{\delta} = 0. \tag{3.3}$$

Solution. Let n be the normal vector to Π and $\langle , \rangle$ the inner product. Then

$$d = \langle f(t_1^0 + ah, t_2^0 + \beta h) - f(t_1^0, t_2^0), n \rangle,$$
$$\delta = |f(t_1^0 + ah, t_2^0 + \beta h) - f(t_1^0, t_2^0)|.$$

Then

$$\begin{aligned}
\lim_{h \to 0} \frac{d}{\delta} &= \lim_{h \to 0} \frac{\langle f(t_1^0 + ah, t_2^0 + \beta h) - f(t_1^0, t_2^0), n \rangle}{|f(t_1^0 + ah, v_0 + \beta h) - f(t_1^0, v_0)|} \\
&= \lim_{h \to 0} \frac{\langle h(af_{t_1}(t_1^0, t_2^0) + \beta f_{t_2}(t_1^0, t_2^0)) + h\varepsilon, n \rangle}{|h(af_{t_1}(t_1^0, t_2^0) + \beta f_{t_2}(t_1^0, t_2^0)) + h\varepsilon|} \\
&= \pm \frac{\langle af_{t_1}(t_1^0, t_2^0) + \beta f_{t_2}(t_1^0, t_2^0), n \rangle}{|af_{t_1}(t_1^0, t_2^0) + \beta f_{t_2}(t_1^0, t_2^0)|}.
\end{aligned}$$

Thus, (3.3) holds if and only if

$$\langle \alpha f_{t_1}(t_1^0, t_2^0) + \beta f_{t_2}(t_1^0, t_2^0), n \rangle = 0. \tag{3.4}$$

1. Let Π be the tangent plane to S at $f(t_1^0, t_2^0)$. Then

$$n \perp f_{t_1}(t_1^0, t_2^0), \quad n \perp f_{t_2}(t_1^0, t_2^0)$$

and (3.4) holds. Hence, (3.3) holds.
2. Let (3.3) hold. Then (3.4) holds. For $\alpha = 1, \beta = 0$, we get

$$\langle f_{t_1}(t_1^0, t_2^0), n \rangle = 0.$$

For $\alpha = 0, \beta = 1$, we find

$$\langle f_{t_2}(t_1^0, t_2^0), n \rangle = 0.$$

Thus,

$$n \perp f_{t_1}(t_1^0, t_2^0), \quad n \perp f_{t_2}(t_1^0, t_2^0)$$

and Π is the tangent plane to S at $f(t_1^0, t_2^0)$. This completes the proof.

Definition 3.14. Let $p \in S$. The straight line passing through p and perpendicular to $T_p S$ is called the normal to the surface S at p.

Let $(U, f = f(t_1, t_2))$ be a surface S, $p = f(t_1^0, t_2^0) \in S$, and

$$f(t_1, t_2) = (f_1(t_1, t_2), f_2(t_1, t_2), f_3(t_1, t_2)), \quad (t_1, t_2) \in U.$$

Then the equations of the normal to S at p are given by

$$\frac{X - f_1(t_1^0, t_2^0)}{\begin{vmatrix} f_{2t_1}(t_1^0, t_2^0) & f_{3t_1}(t_1^0, t_2^0) \\ f_{2t_2}(t_1^0, t_2^0) & f_{3t_2}(t_1^0, t_2^0) \end{vmatrix}} = \frac{Y - f_2(t_1^0, t_2^0)}{\begin{vmatrix} f_{3t_1}(t_1^0, t_2^0) & f_{1t_1}(t_1^0, t_2^0) \\ f_{3t_2}(t_1^0, t_2^0) & f_{1t_2}(t_1^0, t_2^0) \end{vmatrix}} = \frac{Z - f_3(t_1^0, t_2^0)}{\begin{vmatrix} f_{1t_1}(t_1^0, t_2^0) & f_{2t_1}(u_0, v_0) \\ f_{1t_2}(t_1^0, t_2^0) & f_{2t_2}(t_1^0, t_2^0) \end{vmatrix}}.$$

If the surface S is given by

$$z = f(x, y), \quad (x, y) \in U,$$

where $U \subset \mathbb{R}^2$, then the equation of the normal to S at (x_0, y_0, z_0), $(x_0, y_0) \in U$, $z_0 = f(x_0, y_0)$, is given by

$$\frac{x - x_0}{f_x(x_0, y_0)} = \frac{y - y_0}{f_y(x_0, y_0)} = \frac{z - z_0}{-1}.$$

If the surface S is given by

$$F(x,y,z) = 0, \quad (x,y,z) \in V,$$

where $V \subset \mathbb{R}^3$, then the equation of the normal at $(x_0, y_0, z_0) \in V$ is given by

$$\frac{x - x_0}{F_x(x_0, y_0, z_0)} = \frac{y - y_0}{F_y(x_0, y_0, z_0)} = \frac{z - z_0}{F_z(x_0, y_0, z_0)}.$$

Example 3.18. Consider the surface in Example 3.15. Then the equation of the normal at the point $(1, 1, 2)$ is

$$\frac{x - 1}{1} = \frac{y - 1}{1} = \frac{z - 2}{-1}.$$

Example 3.19. Consider the surface in Example 3.16. Then the equation of the normal at the point $(3, 4, -12)$ is

$$\frac{x - 3}{6} = \frac{y - 4}{8} = \frac{z + 12}{-24}$$

or

$$\frac{x - 3}{3} = \frac{y - 4}{4} = \frac{z + 12}{-12}.$$

Example 3.20. Consider the surface in Example 3.17. Then the equation of the normal at the point $(3, 5, 9)$ is

$$\frac{x - 3}{\begin{vmatrix} 2 & 3 \\ 4 & 12 \end{vmatrix}} = \frac{y - 5}{\begin{vmatrix} 3 & 1 \\ 12 & 1 \end{vmatrix}} = \frac{z - 9}{\begin{vmatrix} 1 & 2 \\ 1 & 4 \end{vmatrix}}$$

or

$$\frac{x - 3}{12} = \frac{y - 5}{-9} = \frac{z - 9}{2}.$$

Exercise 3.13. Find the equation of the normal to the surfaces in Exercise 3.11.

Answer 3.3. 1.

$$\frac{x - 1}{2} = \frac{y - 1}{2} = \frac{z - 2}{-1};$$

2.

$$\frac{x + 2}{8} = \frac{y - 1}{8} = z - 4;$$

3.

$$x - 1 = \frac{y - 1}{-2} = z - 1;$$

4.
$$\begin{cases} x = 1, \\ y = 1; \end{cases}$$

5.
$$\frac{x+3}{23} = \frac{y-4}{-19} = \frac{z-17}{5};$$

6.
$$x - 1 = y - 2 = \frac{z-2}{3};$$

7.
$$x - 1 = \frac{y-2}{11} = \frac{z+1}{5};$$

8.
$$x - 1 = \frac{y-2}{-80} = z + 1;$$

9.
$$\frac{x-2}{5} = \frac{y-3}{4} = z - 6;$$

10.
$$x - 2 = \frac{y-1}{2} = \frac{z}{0};$$

11.
$$\frac{x-1}{6} = \frac{y-3}{3} = \frac{z-4}{-2}.$$

Exercise 3.14. Let (x_0, y_0, z_0) be a given point of the surface given by the equation

$$F(x, y, z) = 0.$$

Prove that the gradient vector

$$\operatorname{grad} F(x_0, y_0, z_0) = (F_x(x_0, y_0, z_0), F_y(x_0, y_0, z_0), F_z(x_0, y_0, z_0))$$

is perpendicular to the tangent plane of the surface at this point.

Solution. Let

$$f(t_1, t_2) = (x(t_1, t_2), y(t_1, t_2), z(t_1, t_2))$$

be a local parameterization of the considered surface. Then

$$f_{t_1} = (x_{t_1}, y_{t_1}, z_{t_1}),$$
$$f_{t_2} = (x_{t_2}, y_{t_2}, z_{t_2}),$$

and

$$0 = F_x x_{t_1} + F_y y_{t_1} + F_z z_{t_1}$$
$$= \langle \operatorname{grad} F, f_{t_1} \rangle$$

and

$$0 = F_x x_{t_2} + F_y y_{t_2} + F_z z_{t_2}$$
$$= \langle \operatorname{grad} F, f_{t_2} \rangle.$$

Thus,

$$\operatorname{grad} F \perp f_{t_1}, \quad \operatorname{grad} F \perp f_{t_2}.$$

This completes the solution.

Definition 3.15. An orientation of a surface S is a choice of a normal vector n to $T_p S$.

Definition 3.16. Let S be an oriented surface with orientation $n(p)$. A local parameterization (U, f) of S is said to be compatible if

$$n = \frac{f_{t_1} \times f_{t_2}}{|f_{t_1} \times f_{t_2}|}.$$

3.5 Differentiable maps on a surface

Let $U \subseteq \mathbb{R}^2$, $V \subseteq \mathbb{R}^3$.

Definition 3.17. Suppose that S is a surface in $\mathbb{R}^3$. A map $g : S \to V$ is said to be differentiable if for any local parameterization (U, f) of S the map $g \circ f : U \to V$ is smooth. The map $g \circ f$ is said to be a local representation of g with respect to the local parameterization (U, f) of the surface S.

Example 3.21. Let $S \subseteq V$ be a surface. The inclusion $i : S \to V$ is defined by

$$i(p) = p, \quad p \in S.$$

Note that the inclusion i is a smooth map for any local parameterization (U, f) of S. The parameterization of i is given by

$$i_f = i \circ f = f.$$

Now, suppose that S is a surface with a local parameterization (U, f). Let $g : S \to V$ be a smooth map. We have

$$f(t_1, t_2) = (f_1(t_1, t_2), f_2(t_1, t_2), f_3(t_1, t_2)), \quad (t_1, t_2) \in U,$$

and

$$g(f)(t_1, t_2) = (g_1(f)(t_1, t_2), g_2(f)(t_1, t_2), g_3(f)(t_1, t_2)), \quad (t_1, t_2) \in U.$$

Moreover,

$$g_{lt_j}(f)(t_1, t_2) = g_{lf_1}(f)(t_1, t_2)f_{1t_j}(t_1, t_2) + g_{lf_2}(f)(t_1, t_2)f_{2t_j}(t_1, t_2)$$
$$+ g_{lf_3}(f)(t_1, t_2)f_{3t_j}(t_1, t_2), \quad (t_1, t_2) \in U,$$

with $l \in \{1, 2, 3\}, j \in \{1, 2\}$.

Definition 3.18. Let $S_1, S_2 \subseteq \mathbb{R}^3$ be two surfaces. A map $F : S_1 \to S_2$ is called smooth if the map

$$i \circ F : S_1 \to \mathbb{R}^3$$

is smooth.

Definition 3.19. Let $S_1, S_2 \subseteq \mathbb{R}^3$ be two surfaces. A map $F : S_1 \to S_2$ is said to be a diffeomorphism if F is bijective and F, F^{-1} are smooth maps.

3.6 The differential of a smooth map between two surfaces

Let $I \subseteq \mathbb{R}, V \subseteq \mathbb{R}^3$, and $G : V \to \mathbb{R}^3$ be a smooth map,

$$G(x, y, z) = (g_1(x, y, z), g_2(x, y, z), g_3(x, y, z)), \quad (x, y, z) \in V.$$

Definition 3.20. For any $p = (x_0, y_0, z_0) \in V$, the differential of G at p,

$$d_p G : \mathbb{R}_p^3 \to \mathbb{R}_{G(p)}^3,$$

is a linear map with the matrix

$$\mathcal{G}(x_0, y_0, z_0) = \begin{pmatrix} g_{1x}(x_0, y_0, z_0) & g_{2x}(x_0, y_0, z_0) & g_{3x}(x_0, y_0, z_0) \\ g_{1y}(x_0, y_0, z_0) & g_{2y}(x_0, y_0, z_0) & g_{3y}(x_0, y_0, z_0) \\ g_{1z}(x_0, y_0, z_0) & g_{2x}(x_0, y_0, z_0) & g_{3z}(x_0, y_0, z_0) \end{pmatrix}.$$

Let now

$$f(t) = (f_1(t), f_2(t), f_3(t)), \quad t \in I,$$

be a parameterized curve. Then

$$G \circ f(t) = (g_1(f_1(t), f_2(t), f_3(t)), g_2(f_1(t), f_2(t), f_3(t)), g_3(f_1(t), f_2(t), f_3(t))), \quad t \in I.$$

Hence,

$$(G \circ f)'(t) = (g_{1f_1}(f_1(t), f_2(t), f_3(t))f_1'(t) + g_{1f_2}(f_1(t), f_2(t), f_3(t))f_2'(t)$$
$$+ g_{1f_3}(f_1(t), f_2(t), f_3(t))f_3'(t), g_{2f_1}(f_1(t), f_2(t), f_3(t))f_1'(t)$$

$$+ g_{2f_2}(f_1(t), f_2(t), f_3(t)) f_2'(t) + g_{2f_3}(f_1(t), f_2(t), f_3(t)) f_3'(t),$$
$$g_{3f_1}(f_1(t), f_2(t), f_3(t)) f_1'(t) + g_{3f_2}(f_1(t), f_2(t), f_3(t)) f_2'(t)$$
$$+ g_{3f_3}(f_1(t), f_2(t), f_3(t)) f_3'(t)), \quad t \in I.$$

Thus, $d_p G$ assigns to any tangent vector to $f(t)$ at $t = t_0$ a tangent vector to $G(f)(t)$ at $G(f)(t_0)$.

Definition 3.21. Let S_1 and S_2 be two surfaces, $F : S_1 \to S_2$ be a smooth map between S_1 and S_2, and $p \in S_1$. Then the smooth curve $(I, F \circ f)$ on S_2 corresponds to any smooth curve (I, f) on S_1. If $p = f(t_0)$, $t_0 \in I$, then $F \circ f(t)$ passes through $F(p)$ at $t = t_0$. The map $T_p S_1 \to T_{F(p)} S_2$ assigning to each tangent vector $f(t_0)$ to a parameterized curve $f(t)$ on S, with $f(t_0) = p$, the tangent vector $(F \circ f)(t_0)$ to the parameterized curve $F \circ f$ at $t = t_0$ is said to be the differential of the smooth map $F : S_1 \to S_2$ at the point p.

3.7 The spherical map. The shape operator

Let $S \subseteq \mathbb{R}^3$ be an oriented surface and $\mathbb{S}^2$ be the unit sphere in $\mathbb{R}^3$ centered at the origin $(0, 0, 0)$. Let $p \in S$ and the orientation of S be the unit normal n to S.

Definition 3.22. The map $\Gamma : S \to \mathbb{S}^2$,

$$\Gamma(p) = n(p), \quad p \in S,$$

is said to be the spherical map of the surface S.

Exercise 3.15. Prove that the spherical map $\Gamma : S \to \mathbb{S}^2$ is a smooth map.

Solution. Let (U, f) be a local parameterization of S that is compatible with the orientation of S. Then

$$n(t_1, t_2) = \frac{f_{t_1} \times f_{t_2}}{|f_{t_1} \times f_{t_2}|}.$$

Hence,

$$\Gamma \circ f(t_1, t_2) = \Gamma(f(t_1, t_2))$$
$$= n(t_1, t_2), \quad (t_1, t_2) \in U.$$

Therefore $\Gamma \circ f$ is a smooth map. This completes the solution.

Definition 3.23. The linear operator

$$d_p \Gamma : T_p S \to T_p S$$

is called the shape operator of S at p. It will be denoted by A or A_p.

Below, suppose that (U, f) is a local parameterization of S that is compatible with the orientation of S, and the orientation of S is given by $n(t_1, t_2)$. Then

$$n(t_1, t_2) = \frac{f_{t_1} \times f_{t_2}}{|f_{t_1} \times f_{t_2}|}.$$

Let $v \in T_{f(t_1, t_2)}S$, $v = (v_1, v_2)$ with respect to the orthonormal basis $\{f_{t_1}, f_{t_2}\}$. Then

$$A(v) = v_1 n_{t_1} + v_2 n_{t_2}.$$

In particular, we have

$$A(f_{t_1}) = n_{t_1},$$
$$A(f_{t_2}) = n_{t_2}.$$

Exercise 3.16. Prove that the shape operator A is a self-adjoint, i. e.,

$$A(v) \cdot w = w \cdot A(v),$$

for any $v, w \in T_p S$.

Solution. Consider the equations

$$n \cdot f_{t_1} = 0,$$
$$n \cdot f_{t_2} = 0,$$

where we differentiate with respect to t_2 and t_1, respectively, and get

$$0 = n_{t_2} \cdot f_{t_1} + n \cdot f_{t_1 t_2},$$
$$0 = n_{t_1} \cdot f_{t_2} + n \cdot f_{t_1 t_2}.$$

Using the latter two equations, we conclude that

$$n_{t_2} \cdot f_{t_1} = n_{t_1} \cdot f_{t_2}.$$

Let $v = (v_1, v_2)$, $w = (w_1, w_2) \in T_p S$ with respect to the basis $\{f_{t_1}, f_{t_2}\}$. Then

$$v = v_1 f_{t_1} + v_2 f_{t_2},$$
$$w = w_1 f_{t_1} + w_2 f_{t_2},$$
$$A(v) = v_1 n_{t_1} + v_2 n_{t_2},$$
$$A(w) = w_1 n_{t_1} + w_2 n_{t_2}.$$

Hence,

$$A(v) \cdot w = (v_1 n_{t_1} + v_2 n_{t_2}) \cdot (w_1 f_{t_1} + w_2 f_{t_1})$$
$$= (v_1 n_{t_1}) \cdot (w_1 f_{t_1}) + (v_1 n_{t_1}) \cdot (w_2 f_{t_1}) + (v_2 n_{t_2}) \cdot (w_1 f_{t_1}) + (v_2 n_{t_2}) \cdot (w_1 f_{t_2})$$
$$= (v_1 w_1)(n_{t_1} \cdot f_{t_1}) + (v_1 w_2)(n_{t_1} \cdot f_{t_2}) + (v_2 w_1)(n_{t_2} \cdot f_{t_1}) + (v_2 w_2)(n_{t_2} \cdot f_{t_2})$$
$$= (v_1 w_1)(f_{t_1} \cdot n_{t_1}) + (v_1 w_2)(f_{t_1} \cdot n_{t_2}) + (v_2 w_1)(f_{t_2} \cdot n_{t_1}) + (v_2 w_2)(f_{t_1} \cdot n_{t_1})$$
$$= (v_1 f_{t_1}) \cdot (w_1 n_{t_1}) + (v_1 f_{t_1}) \cdot (w_2 n_{t_2}) + (v_2 f_{t_2}) \cdot (w_1 n_{t_1}) + (w_2 f_{t_1}) \cdot (w_2 n_{t_2})$$
$$= (v_1 f_{t_1}) \cdot A(w) + (v_2 f_{t_2}) \cdot A(w)$$
$$= v \cdot A(w).$$

This completes the solution.

Exercise 3.17. Prove that for each tangent space $T_p S$ there is an orthonormal basis of the eigenvectors of the shape operator A.

Solution. Let $\lambda_1, \lambda_2 \in \mathbb{R}$, $\lambda_1 \neq \lambda_2$, be eigenvalues of the operator A that correspond to eigenvectors v and w, respectively. Then

$$A(v) = \lambda_1 v,$$
$$A(w) = \lambda_2 w.$$

Hence,

$$A(v) \cdot w = (\lambda_1 v) \cdot w = \lambda_1 (v \cdot w)$$

and

$$v \cdot A(w) = v \cdot (\lambda_2 w) = \lambda_2 (v \cdot w).$$

Now, by Exercise 3.16, we have that

$$A(v) \cdot w = v \cdot A(w).$$

From here, we find

$$\lambda_1 (v \cdot w) = \lambda_2 (v \cdot w),$$

hence

$$(\lambda_1 - \lambda_2)(v \cdot w) = 0$$

and

$$v \cdot w = 0.$$

Let

$$\tilde{v} = \frac{v}{|v|},$$

$$\tilde{w} = \frac{w}{|w|}.$$

Then

$$|\tilde{v}| = 1, \quad |\tilde{w}| = 1$$

and

$$\tilde{v} \cdot \tilde{w} = 0.$$

This completes the solution.

3.8 The first fundamental form of a surface

Suppose that S is an oriented surface.

Definition 3.24. For any $p \in S$ and $v, w \in T_p S$, the first fundamental form of S at p is defined by

$$\phi_1(v, w) = v \cdot w.$$

Let (U, f) be a local parameterization of S that is compatible with the orientation of S. Let also $v = (v_1, v_2)$, $w = (w_1, w_2) \in T_p S$ with respect to the basis $\{f_{t_1}, f_{t_2}\}$. Then

$$v = v_1 f_{t_1} + v_2 f_{t_2},$$
$$w = w_1 f_{t_1} + w_2 f_{t_2}.$$

Thus

$$\phi_1(v, w) = v \cdot w$$
$$= (v_1 f_{t_1} + v_2 f_{t_2}) \cdot (w_1 f_{t_1} + w_2 f_{t_2})$$
$$= (v_1 w_1)(f_{t_1} \cdot f_{t_1}) + (v_1 w_2)(f_{t_1} \cdot f_{t_2}) + (v_2 w_1)(f_{t_1} \cdot f_{t_2}) + (v_2 w_2)(f_{t_2} \cdot f_{t_2}).$$

Define

$$E(t_1, t_2) = f_{t_1} \cdot f_{t_1},$$
$$F(t_1, t_2) = f_{t_1} \cdot f_{t_2},$$
$$G(t_1, t_2) = f_{t_1} \cdot f_{t_2}, \quad (t_1, t_2) \in U,$$

and the matrix

$$\mathcal{G}(t_1, t_2) = \begin{pmatrix} E(t_1, t_2) & F(t_1, t_2) \\ F(t_1, t_2) & G(t_1, t_2) \end{pmatrix}, \quad (t_1, t_2) \in U.$$

Then

$$\begin{aligned} \phi_1(v, w) &= (v_1 w_1)E(t_1, t_2) + (v_1 w_2)F(t_1, t_2) + (v_2 w_1)F(t_1, t_2) + (v_2 w_2)G(t_1, t_2) \\ &= v_1(E(t_1, t_2)w_1 + F(t_1, t_2)w_2) + v_2(F(t_1, t_2)w_1 + G(t_1, t_2)w_2) \\ &= v\mathcal{G}(t_1, t_2)w. \end{aligned}$$

Definition 3.25. The matrix $\mathcal{G}$ is said to be the matrix of the first fundamental form for S.

Example 3.22. We will find the matrix of the first fundamental form for the surface

$$f(t_1, t_2) = (\cos t_1 \cos t_2, \cos t_1 \sin t_2, \sin t_1), \quad (t_1, t_2) \in [0, 2\pi] \times [0, 2\pi].$$

Here

$$\begin{aligned} f_1(t_1, t_2) &= \cos t_1 \cos t_2, \\ f_2(t_1, t_2) &= \cos t_1 \sin t_2, \\ f_3(t_1, t_2) &= \sin t_1, \quad (t_1, t_2) \in [0, 2\pi] \times [0, 2\pi]. \end{aligned}$$

Then

$$\begin{aligned} f_{1t_1}(t_1, t_2) &= -\sin t_1 \cos t_2, \\ f_{2t_1}(t_1, t_2) &= -\sin t_1 \sin t_2, \\ f_{3t_1}(t_1, t_2) &= \cos t_1, \\ f_{1t_2}(t_1, t_2) &= -\cos t_1 \sin t_2, \\ f_{2t_2}(t_1, t_2) &= \cos t_1 \cos t_2, \\ f_{3t_2}(t_1, t_2) &= 0, \quad (t_1, t_2) \in [0, 2\pi] \times [0, 2\pi], \end{aligned}$$

and

$$\begin{aligned} f_{t_1}(t_1, t_2) &= (f_{1t_1}(t_1, t_2), f_{2t_1}(t_1, t_2), f_{3t_1}(t_1, t_2)) \\ &= (-\sin t_1 \cos t_2, -\sin t_1 \sin t_2, \cos t_1), \\ f_{t_2}(t_1, t_2) &= (f_{1t_2}(u, t_2), f_{2t_2}(u, t_2), f_{3t_2}(u, t_2)) \\ &= (-\cos t_1 \sin t_2, \cos t_1 \cos t_2, 0), \quad (t_1, t_2) \in [0, 2\pi] \times [0, 2\pi]. \end{aligned}$$

Hence,

$$\begin{aligned} f_{t_1}(t_1, t_2) \cdot f_{t_1}(t_1, t_2) &= (-\sin t_1 \cos t_2)^2 + (-\sin t_1 \sin t_2)^2 + (\cos t_1)^2 \\ &= (\sin t_1)^2 (\cos t_2)^2 + (\sin t_1)^2 (\sin t_2)^2 + (\cos t_1)^2 \\ &= (\sin t_1)^2 + (\cos t_2)^2 \end{aligned}$$

$$= 1,$$

$$f_{t_1}(t_1, t_2) \cdot f_{t_2}(t_1, t_2) = \sin t_1 \cos t_1 \sin t_2 \cos t_2 - \sin t_1 \cos t_1 \sin t_2 \cos t_2$$

$$= 0,$$

$$f_{t_2}(t_1, t_2) \cdot f_{t_2}(t_1, t_2) = (-\cos t_1 \sin t_2)^2 + (\cos t_1 \cos t_2)^2$$

$$= (\cos t_1)^2 (\sin t_2)^2 + (\cos t_1)^2 (\cos t_2)^2$$

$$= (\cos t_1)^2, \quad (t_1, t_2) \in [0, 2\pi] \times [0, 2\pi].$$

Therefore

$$G(t_1, t_2) = \begin{pmatrix} f_{t_1}(t_1, t_2) \cdot f_{t_1}(t_1, t_2) & f_{t_1}(t_1, t_2) \cdot f_v(t_1, t_2) \\ f_{t_1}(t_1, t_2) \cdot f_{t_2}(t_1, t_2) & f_{t_2}(t_1, t_2) \cdot f_{t_2}(t_1, t_2) \end{pmatrix}$$

$$= \begin{pmatrix} 1 & 0 \\ 0 & (\cos t_1)^2 \end{pmatrix}, \quad (t_1, t_2) \in [0, 2\pi] \times [0, 2\pi].$$

Exercise 3.18. Let $a, c \in \mathbb{R}$. Find the matrix of the first fundamental form for the following surfaces:

1.
$$(a \cos t_1 \cos t_2, a \cos t_1 \sin t_2, c \sin t_1), \quad (t_1, t_2) \in [0, 2\pi] \times [0, 2\pi];$$

2.
$$(a \cosh t_1 \cos t_2, a \cosh t_1 \sin t_2, c \sinh t_1), \quad (t_1, t_2) \in \mathbb{R}^2;$$

3.
$$(a \sinh t_1 \cos t_2, a \sinh t_1 \sin t_2, c \cosh t_1), \quad (t_1, t_2) \in \mathbb{R}^2;$$

4.
$$(t_1 \cos t_2, t_1 \sin t_2, t_1^2), \quad t_1 \in \mathbb{R}, \quad t_2 \in [0, 2\pi];$$

5.
$$(R \cos t_2, R \sin t_2, t_1), \quad t_1 \in \mathbb{R}, \quad t_2 \in [0, 2\pi],$$

where $R \in \mathbb{R}$.

Answer 3.4. 1.
$$\begin{pmatrix} a^2 (\sin t_1)^2 + c^2 (\cos t_1)^2 & 0 \\ 0 & a^2 (\cos t_1)^2 \end{pmatrix}, \quad (t_1, t_2) \in [0, 2\pi] \times [0, 2\pi];$$

2.
$$\begin{pmatrix} a^2 (\sinh t_1)^2 + c^2 (\cosh t_1)^2 & 0 \\ 0 & a^2 (\cosh t_1)^2 \end{pmatrix}, \quad (t_1, t_2) \in \mathbb{R}^2;$$

3.
$$\begin{pmatrix} a^2 (\cosh t_1)^2 + c^2 (\sinh t_1)^2 & 0 \\ 0 & a^2 (\sinh t_1)^2 \end{pmatrix}, \quad (t_1, t_2) \in \mathbb{R}^2;$$

4.
$$\begin{pmatrix} 1 + 4t_1^2 & 0 \\ 0 & t_1^2 \end{pmatrix}, \quad (t_1, t_2) \in \mathbb{R}^2;$$

5.
$$\begin{pmatrix} 1 & 0 \\ 0 & R^2 \end{pmatrix}, \quad t_1 \in \mathbb{R}, \quad t_1 \in [0, 2\pi].$$

3.9 Applications of the first fundamental form

Let S be an oriented surface with a local parameterization (U, f) that is compatible with the orientation of S.

3.9.1 The length of a segment of a curve on a surface

Let $(I, g = g(s))$ be a parameterized curve on S, $g(s) \subset f(U)$. Let also

$$g(s) = (t_1(s), t_2(s)), \quad t \in I.$$

Then

$$g'(s) = (t_1'(s), t_2'(s)), \quad t \in I,$$

and

$$g'(s) \cdot g'(s) = g'(s)\mathcal{G}(t_1(s), t_2(s))g'(s)$$

$$= (t_1'(s), t_2'(s)) \begin{pmatrix} E(t_1(s), t_2(s)) & F(t_1(s), t_2(s)) \\ F(t_1(s), t_2(s)) & G(t_1(s), t_2(s)) \end{pmatrix} \begin{pmatrix} t_1(s) \\ t_2(s) \end{pmatrix}$$

$$= (t_1'(s), t_2'(s)) \begin{pmatrix} E(t_1(s), t_2(s))t_1'(s) + F(t_1(s), t_2(s))t_2'(s) \\ F(t_1(s), t_2(s))t_1'(s) + G(t_1(s), t_2(s))t_2'(s) \end{pmatrix}$$

$$= (t_1'(s))^2 E(t_1(s), t_2(s)) + 2F(t_1(s), t_2(s))t_1'(s)t_2'(s)$$

$$+ (t_2'(s))^2 G(t_1(s), t_2(s)), \quad t \in I.$$

Definition 3.26. The length of the segment of the curve $g(s)$ between s_1 and s_2 on the surface S is defined by

$$l(s_1, s_2) = \int_{s_1}^{s_2} |g'(s)|^{\frac{1}{2}} ds.$$

We have

$$l(s_1, s_2) = \int_{s_1}^{s_2} ((t_1'(s))^2 E(t_1(s), t_2(s)) + 2F(t_1(s), t_2(s))t_1'(s)t_2'(s)$$

$$+ (t_2'(s))^2 G(t_1(s), t_2(s)))^{\frac{1}{2}} ds.$$

Example 3.23. Consider the sphere in Example 3.22 and the curve given by the local equations

$$t_1 = 0,$$
$$t_2 = s, \quad s \in [0, 2\pi].$$

Using the computations in Example 3.22, we find

$$\mathcal{G}(t_1(s), t_2(s)) = \begin{pmatrix} 1 & 0 \\ 0 & (\cos t_1)^2 \end{pmatrix} = \begin{pmatrix} 1 & 0 \\ 0 & 1 \end{pmatrix}.$$

Next,

$$t_1'(s) = 0,$$
$$t_2'(s) = 1, \quad s \in [0, 2\pi].$$

Then

$$l\left(\frac{\pi}{2}, \pi\right) = \int_{\frac{\pi}{2}}^{\pi} \sqrt{1 \cdot 0^2 + 2 \cdot 0 \cdot 0 \cdot 1 + 1 \cdot 1^2} \, dt$$

$$= \int_{\frac{\pi}{2}}^{\pi} dt$$

$$= \frac{\pi}{2}.$$

Exercise 3.19. Consider the surface (see Fig. 3.16)

$$(t_1^2 + t_2^2, t_1^2 - t_2^2, t_1 t_2), \quad (t_1, t_2) \in \mathbb{R}^2.$$

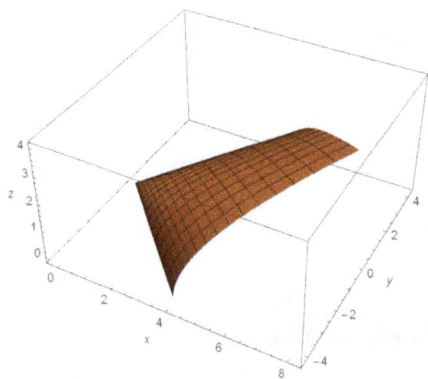

Figure 3.16: The surface $(t_1^2 + t_2^2, t_1^2 - t_2^2, t_1 t_2)$, $t_1, t_2 \in [0, 2]$.

1. Find the matrix of the first fundamental form.
2. Find the length of the segment of the curve between s_1 and s_2

$$t_1 = 1.$$

3. Find the length of the segment of the curve between s_1 and s_2

$$t_2 = 1.$$

4. Find the length of the segment of the curve between s_1 and s_2

$$t_2 = at_1,$$

where $a \in \mathbb{R}$.

Answer 3.5. 1.

$$\begin{pmatrix} 8t_1^2 + t_2^2 & t_1 t_2 \\ t_1 t_2 & 8t_2^2 + t_1^2 \end{pmatrix};$$

2.

$$\frac{1}{24}\left((8t_2^2(s_2) + 1)^{3/2} - (8t_2^2(s_1) + 1)^{3/2}\right);$$

3.

$$\frac{1}{24}\left((8t_2^2(s_2) + 1)^{3/2} - (8t_2^2(s_1) + 1)^{3/2}\right);$$

4.

$$\sqrt{2a^4 + a^2 + 2(t_1^2(s_2) - t_1^2(s_1))}.$$

3.9.2 The angle between two curves on a surface

Let $(I, g_1 = g_1(t))$ and $(J, g_2 = g_2(s))$ be two parameterized curves on S such that

$$g_1(t_0) = g_2(s_0) = f(u_0, v_0).$$

Let also

$$g_1(t) = (u_1(t), v_1(t)), \quad t \in I,$$
$$g_2(s) = (u_2(s), v_2(s)), \quad s \in J.$$

Then

$$g_1'(t) = (u_1'(t), v_1'(t)), \quad t \in I,$$
$$g_2'(s) = (u_2'(s), v_2'(s)), \quad s \in J.$$

Definition 3.27. The angle θ between the curves g_1 and g_2 at the point $f(u_0, v_0)$ is defined as follows:

$$\cos\theta = \frac{g_1'(t_0) \cdot g_2'(s_0)}{|g_1'(t_0)||g_2'(s_0)|}.$$

We have

$$g_1'(t_0) \cdot g_2'(s_0) = (u_1'(t_0), v_1'(t_0)) \begin{pmatrix} E(u_0, v_0) & F(u_0, v_0) \\ F(u_0, v_0) & G(u_0, v_0) \end{pmatrix} \begin{pmatrix} u_2'(s_0) \\ v_2'(s_0) \end{pmatrix}$$

$$= (u_1'(t_0), v_1'(t_0)) \begin{pmatrix} E(u_0, v_0)u_2'(s_0) + F(u_0, v_0)v_2'(s_0) \\ F(u_0, v_0)u_2'(s_0) + G(u_0, v_0)v_2'(s_0) \end{pmatrix}$$

$$= (u_1'(t_0)u_2'(s_0))E(u_0, v_0)$$
$$+ (u_1'(t_0)v_2'(s_0) + v_1'(t_0)u_2'(s_0))F(u_0, v_0)$$
$$+ (v_1'(t_0)v_2'(s_0))G(u_0, v_0)$$

and

$$g_1'(t_0) \cdot g_1'(t_0) = (u_1'(t_0), v_1'(t_0)) \begin{pmatrix} E(u_0, v_0) & F(u_0, v_0) \\ F(u_0, v_0) & G(u_0, v_0) \end{pmatrix} \begin{pmatrix} u_1'(t_0) \\ v_1'(t_0) \end{pmatrix}$$

$$= (u_1'(t_0), v_1'(t_0)) \begin{pmatrix} E(u_0, v_0)u_1'(t_0) + F(u_0, v_0)v_1'(t_0) \\ F(u_0, v_0)u_1'(t_0) + G(u_0, v_0)v_1'(t_0) \end{pmatrix}$$

$$= (u_1'(t_0))^2 E(u_0, v_0) + 2(u_1'(t_0)v_1'(t_0))F(u_0, v_0)$$
$$+ (v_1'(t_0))^2 G(u_0, v_0),$$

and

$$g_2'(s_0) \cdot g_2'(s_0) = (u_2'(s_0), v_2'(s_0)) \begin{pmatrix} E(u_0, v_0) & F(u_0, v_0) \\ F(u_0, v_0) & G(u_0, v_0) \end{pmatrix} \begin{pmatrix} u_2'(s_0) \\ v_2'(s_0) \end{pmatrix}$$

$$= (u_2'(s_0), v_2'(s_0)) \begin{pmatrix} E(u_0, v_0)u_2(s_0) + F(u_0, v_0)v_2(s_0) \\ F(u_0, v_0)u_2(s_0) + G(u_0, v_0)v_2(s_0) \end{pmatrix}$$

$$= (u_2'(s_0))^2 E(u_0, v_0) + 2(u_2'(s_0)v_2'(s_0))F(u_0, v_0)$$
$$+ (v_2'(s_0))^2 G(u_0, v_0).$$

Example 3.24. We will find the angle between the lines

$$u + v = 0 \quad \text{and} \quad u - v = 0$$

on the helicoid (see Fig. 3.17)

$$f(u, v) = (u \cos v, u \sin v, av), \quad u \in \mathbb{R}, \quad v \in [0, 2\pi],$$

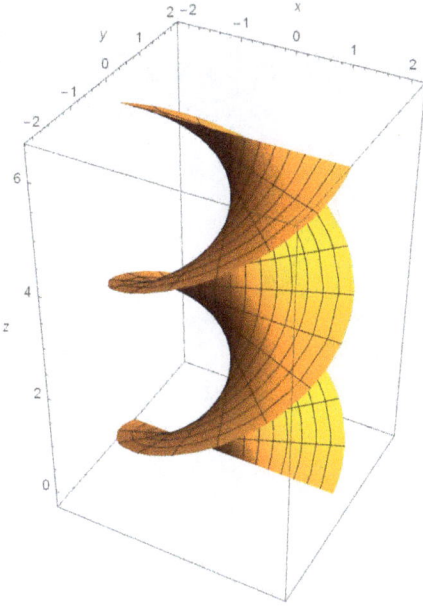

Figure 3.17: The helicoid $(u \cos v, u \sin v, v)$, $u \in [-2, 2]$, $v \in [0, 2\pi]$.

where $a \in \mathbb{R}$ is a given parameter. We have

$$f_u(u, v) = (\cos v, \sin v, 0),$$
$$f_v(u, v) = (-u \sin v, u \cos v, a), \quad u \in \mathbb{R}, \quad v \in [0, 2\pi].$$

Then

$$f_u(u, v) \cdot f_u(u, v) = (\cos v)^2 + (\sin v)^2 = 1,$$
$$f_u(u, v) \cdot f_v(u, v) = -u \sin v \cos v + u \sin v \cos v = 0,$$
$$f_v(u, v) \cdot f_v(u, v) = (-u \sin v)^2 + (u \cos v)^2 + a^2$$
$$= u^2(\sin v)^2 + u^2(\cos v)^2 + a^2$$
$$= u^2 + a^2, \quad u \in \mathbb{R}, \quad v \in [0, 2\pi],$$

and the matrix of the first fundamental form of the considered helicoid is as follows:

$$\mathcal{G}(u, v) = \begin{pmatrix} 1 & 0 \\ 0 & u^2 + a^2 \end{pmatrix}, \quad u \in \mathbb{R}, \quad v \in [0, 2\pi].$$

Note that

$$t_0 = s_0 = u_0 = v_0 = 0$$

and

$$g_1(t) = (t, -t), \quad t \in \mathbb{R},$$
$$g_2(s) = (s, s), \quad s \in \mathbb{R}.$$

Hence,

$$g_1'(t) = (1, -1), \quad t \in \mathbb{R},$$
$$g_2'(s) = (1, 1), \quad s \in \mathbb{R},$$

and

$$g_1'(t) \cdot g_2'(s) = (1, -1) \begin{pmatrix} 1 & 0 \\ 0 & u^2 + a^2 \end{pmatrix} \begin{pmatrix} 1 \\ -1 \end{pmatrix}$$

$$= (1, -1) \begin{pmatrix} 1 \\ -u^2 - a^2 \end{pmatrix}$$

$$= 1 + u^2 + a^2,$$

$$g_1'(t) \cdot g_2'(s) = (1, -1) \begin{pmatrix} 1 & 0 \\ 0 & u^2 + a^2 \end{pmatrix} \begin{pmatrix} 1 \\ 1 \end{pmatrix}$$

$$= (1, -1) \begin{pmatrix} 1 \\ u^2 + a^2 \end{pmatrix}$$

$$= 1 - u^2 - a^2,$$

$$g_2'(s) \cdot g_2'(s) = (1, 1) \begin{pmatrix} 1 & 0 \\ 1 & u^2 + a^2 \end{pmatrix} \begin{pmatrix} 1 \\ 1 \end{pmatrix}$$

$$= (1, 1) \begin{pmatrix} 1 \\ u^2 + a^2 \end{pmatrix}$$

$$= 1 + u^2 + a^2, \quad t, s, u \in \mathbb{R}, \quad v \in [0, 2\pi].$$

Consequently,

$$\cos \theta = \frac{g_1'(0) \cdot g_2'(0)}{|g_1'(0)||g_2'(0)|}$$

$$= \frac{1 - a^2}{\sqrt{1 + a^2}\sqrt{1 + a^2}}$$

$$= \frac{1 - a^2}{1 + a^2}.$$

Exercise 3.20. Find the angle between the curves

$$v = 2u \quad \text{and} \quad v = -2u$$

on the surface with a matrix of its first fundamental form given by

$$\begin{pmatrix} 1 & 0 \\ 0 & 1 \end{pmatrix}.$$

Answer 3.6.

$$\cos\theta = -\frac{3}{5}.$$

3.9.3 The area of a parameterized surface

Definition 3.28. The area of a surface S is defined as follows:

$$A = \iint_U \left(E(t_1, t_2) G(t_1, t_2) - (F(t_1, t_2))^2 \right)^{\frac{1}{2}} dt_1 dt_2.$$

Example 3.25. We will find the area of the surface in Example 3.22. We have

$$A = \int_0^{2\pi} \sqrt{(\cos t_1)^2} dt_1$$

$$= \int_0^{2\pi} |\cos t_1| dt_1$$

$$= \int_0^{\frac{\pi}{2}} \cos t_1 dt_1 - \int_{\frac{\pi}{2}}^{\pi} \cos t_1 dt_1 - \int_{\pi}^{\frac{3\pi}{2}} \cos t_1 dt_1 + \int_{\frac{3\pi}{2}}^{\pi} \cos t_1 dt_1$$

$$= \sin t_1 \big|_{t_1=0}^{t_1=\frac{\pi}{2}} - \sin t_1 \big|_{t_1=\frac{\pi}{2}}^{t_1=\pi} - \sin t_1 \big|_{t_1=\pi}^{t_1=\frac{3\pi}{2}} + \sin t_1 \big|_{t_1=\frac{3\pi}{2}}^{t_1=2\pi}$$

$$= 1 + 1 + 1 + 1$$

$$= 4.$$

Exercise 3.21. Find the area of the rectangle bounded by the helicoid

$$f(t_1, t_2) = (t_1 \cos t_2, t_1 \sin t_2, at_2), \quad (t_1, t_2) \in \mathbb{R}^2,$$

and the lines

$$t_1 = 0, \quad t_1 = a, \quad t_2 = 0, \quad t_2 = 1.$$

Here $a \in \mathbb{R}$ is a given parameter.

Answer 3.7.

$$\frac{a}{2}\left(\sqrt{2} + \log(1 + \sqrt{2})\right).$$

3.10 The matrix of the shape operator

Let (U, f) be a local parameterization of S. Let also $\{f_{t_1}, f_{t_2}\}$ be the basis. By $\mathcal{A}$ we will denote the matrix of the shape operator A. We have

$$A(f_{t_1}) = n_{t_1},$$
$$A(f_{t_2}) = n_{t_2}.$$

Then

$$(n_{t_1}, n_{t_2}) = (f_{t_1}, f_{t_2})\mathcal{A}. \tag{3.5}$$

Note that

$$\begin{pmatrix} f_{t_1} \\ f_{t_2} \end{pmatrix} (n_{t_1}, n_{t_2}) = \begin{pmatrix} f_{t_1} \cdot n_{t_1} & f_{t_1} \cdot n_{t_2} \\ f_{t_2} \cdot n_{t_1} & f_{t_2} \cdot n_{t_2} \end{pmatrix}.$$

Let

$$L = -f_{t_1} \cdot n_{t_1},$$
$$M = -f_{t_1} \cdot n_{t_2} = -f_{t_2} \cdot n_{t_1},$$
$$N = -f_{t_2} \cdot n_{t_2}.$$

Set

$$\mathcal{H} = \begin{pmatrix} L & M \\ M & N \end{pmatrix}.$$

Therefore

$$\begin{pmatrix} f_{t_1} \\ f_{t_2} \end{pmatrix} (n_{t_1}, n_{t_2}) = -\mathcal{H}.$$

Hence using (3.5), we get

$$\mathcal{H} = \begin{pmatrix} L & M \\ M & N \end{pmatrix}$$

$$= -\begin{pmatrix} f_{t_1} \cdot n_{t_1} & f_{t_1} \cdot n_{t_2} \\ f_{t_2} \cdot n_{t_1} & f_{t_2} \cdot n_{t_2} \end{pmatrix}$$

$$= -\begin{pmatrix} f_{t_1} \\ f_{t_2} \end{pmatrix} (n_{t_1}, n_{t_2})$$

$$= -\begin{pmatrix} f_{t_1} \\ f_{t_2} \end{pmatrix} (f_{t_1}, f_{t_2})\mathcal{A}$$

$$= - \begin{pmatrix} f_{t_1} \cdot f_{t_1} & f_{t_1} \cdot f_{t_2} \\ f_{t_1} \cdot f_{t_2} & f_{t_2} \cdot f_{t_2} \end{pmatrix} \mathcal{A}$$

$$= - \begin{pmatrix} E & F \\ F & G \end{pmatrix} \mathcal{A}$$

$$= -\mathcal{G}\mathcal{A},$$

whereupon

$$\mathcal{A} = -\mathcal{G}^{-1}\mathcal{H}.$$

Note that

$$\mathcal{G}^{-1} = \frac{1}{EG - F^2} \begin{pmatrix} G & -F \\ -F & E \end{pmatrix}.$$

Thus,

$$\mathcal{A} = -\frac{1}{EG - F^2} \begin{pmatrix} G & -F \\ -F & E \end{pmatrix} \begin{pmatrix} L & M \\ M & N \end{pmatrix}$$

$$= -\frac{1}{EG - F^2} \begin{pmatrix} GL - FM & GM - FN \\ EM - FL & EM - FN \end{pmatrix}.$$

Now, we differentiate the following equation:

$$f_{t_1} \cdot n = 0 \tag{3.6}$$

with respect to t_1 and obtain

$$0 = f_{t_1 t_1} \cdot n + f_{t_1} \cdot n_{t_1},$$

whereupon

$$L = n \cdot f_{t_1 t_1}.$$

We differentiate equation (3.6) with respect to t_2 and arrive at

$$0 = f_{t_1 t_2} \cdot n + f_{t_1} \cdot n_{t_2},$$

from where

$$M = f_{t_1 t_2} \cdot n.$$

We differentiate the following equation:

$$f_{t_2} \cdot n = 0$$

with respect to t_2 and find

$$0 = f_{t_2 t_2} \cdot n + f_{t_2} \cdot n_{t_2},$$

whereupon

$$N = n \cdot f_{t_2 t_2}.$$

Note that

$$(f_{t_1} \times f_{t_2}) \cdot (f_{t_1} \times f_{t_2}) = (f_{t_1} \cdot f_{t_1})(f_{t_2} \cdot f_{t_2}) - (f_{t_1} \cdot f_{t_2})(f_{t_1} \cdot f_{t_2})$$
$$= EG - F^2$$

and

$$|f_{t_1} \times f_{t_2}| = (EG - F^2)^{\frac{1}{2}}.$$

Let

$$W = (EG - F^2)^{\frac{1}{2}}.$$

Then

$$|f_{t_1} \times f_{t_2}| = W$$

and

$$L = \frac{1}{W}((f_{t_1} \times f_{t_2}) \cdot f_{t_1 t_1}),$$
$$M = \frac{1}{W}((f_{t_1} \times f_{t_2}) \cdot f_{t_1 t_2}),$$
$$N = \frac{1}{W}((f_{t_1} \times f_{t_2}) \cdot f_{t_2 t_2}).$$

Example 3.26. Consider the surface (see Fig. 3.18)

$$f(t_1, t_2) = (t_1 + t_2, t_1^2 + t_2^2, t_1 t_2), \quad (t_1, t_2) \in \mathbb{R}^2.$$

Then

$$f_{t_1}(t_1, t_2) = (1, 2t_1, t_2),$$
$$f_{t_2}(t_1, t_2) = (1, 2t_2, t_1),$$
$$f_{t_1 t_1}(t_1, t_2) = (0, 2, 0),$$
$$f_{t_1 t_2}(t_1, t_2) = (0, 0, 1),$$
$$f_{t_2 t_2}(t_1, t_2) = (0, 2, 0), \quad (t_1, t_2) \in \mathbb{R}^2,$$

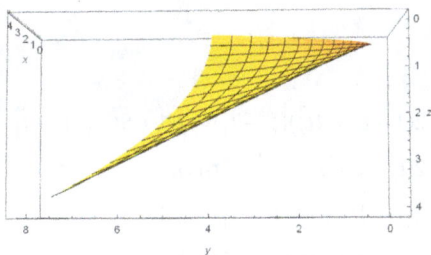

Figure 3.18: The surface $f(t_1, t_2) = (t_1 + t_2, t_1^2 + t_2^2, t_1 t_2)$, $(t_1, t_2) \in [0, 2]$.

and

$$E(t_1, t_2) = f_{t_1}(t_1, t_2) \cdot f_{t_1}(t_1, t_2)$$
$$= 1 + 4t_1^2 + t_2^2,$$
$$F(t_1, t_2) = f_{t_1}(t_1, t_2) \cdot f_{t_2}(t_1, t_2)$$
$$= 1 + 4t_1 t_2 + t_1 t_2$$
$$= 1 + 5t_1 t_2,$$
$$G(t_1, t_2) = f_{t_2}(t_1, t_2) \cdot f_{t_2}(t_1, t_2)$$
$$= 1 + t_1^2 + 4t_2^2, \quad (t_1, t_2) \in \mathbb{R}^2,$$

and

$$W(t_1, t_2) = \left(E(t_1, t_2) G(t_1, t_2) - \left(F(t_1, t_2) \right)^2 \right)^{\frac{1}{2}}$$
$$= \left((1 + 4t_1^2 + t_2^2)(1 + t_1^2 + 4t_2^2) - (1 + 5t_1 t_2)^2 \right)^{\frac{1}{2}}$$
$$= \left(1 + t_1^2 + 4t_2^2 + 4t_1^2 + 4t_1^4 + 16t_1^2 t_2^2 + t_2^2 + t_1^2 t_2^2 + 4t_2^4 \right.$$
$$\left. - 1 - 10t_1 t_2 - 25t_1^2 t_2^2 \right)^{\frac{1}{2}}$$
$$= \left(5t_1^2 + 5t_2^2 + 4t_1^4 + 4t_2^4 - 10t_1 t_2 - 25t_1^2 t_2^2 \right)^{\frac{1}{2}},$$
$$= |t_1 - t_2| \left(4(t_1 + t_2)^2 + 5 \right)^{1/2} \quad (t_1, t_2) \in \mathbb{R}^2,$$

and

$$f_{t_1}(t_1, t_2) \times f_{t_2}(t_1, t_2) = (2(t_1^2 - t_2^2), t_2 - t_1, 2(t_2 - t_1)),$$
$$W(t_1, t_2) L(t_1, t_2) = \left(f_{t_1}(t_1, t_2) \times f_{t_2}(t_1, t_2) \right) \cdot f_{t_1 t_1}(t_1, t_2)$$
$$= 2(t_2 - t_1),$$
$$W(t_1, t_2) M(t_1, t_2) = \left(f_{t_1}(t_1, t_2) \times f_{t_2}(t_1, t_2) \right) \cdot f_{t_1 t_2}(t_1, t_2)$$
$$= 2(t_2 - t_1),$$
$$W(t_1, t_2) N(t_1, t_2) = \left(f_{t_1}(t_1, t_2) \times f_{t_2}(t_1, t_2) \right) \cdot f_{t_2 t_2}(t_1, t_2)$$
$$= 2(t_2 - t_1), \quad (t_1, t_2) \in \mathbb{R}^2,$$

and

$$W(t_1,t_2)(G(t_1,t_2)L(t_1,t_2) - F(t_1,t_2)M(t_1,t_2)) = 2(1+t_1^2+4t_2^2)(t_2-t_1) - 2(1+5t_1t_2)(t_2-t_1)$$
$$= 2(t_2-t_1)(4t_2^2+t_1^2-5t_1t_2),$$
$$W(t_1,t_2)(G(t_1,t_2)M(t_1,t_2) - F(t_1,t_2)N(t_1,t_2)) = 2(1+4t_2^2+t_1^2)(t_2-t_1) - 2(1+5t_1t_2)(t_2-t_1)$$
$$= 2(t_2-t_1)(4t_2^2+t_1^2-5t_1t_2),$$
$$W(t_1,t_2)(E(t_1,t_2)M(t_1,t_2) - F(t_1,t_2)L(t_1,t_2)) = 2(1+4t_1^2+t_2^2)(t_2-t_1) - 2(1+5t_1t_2)(t_2-t_1)$$
$$= 2(t_2-t_1)(4t_1^2+t_2^2-5t_1t_2),$$
$$W(t_1,t_2)(E(t_1,t_2)M(t_1,t_2) - F(t_1,t_2)N(t_1,t_2)) = 2(1+4t_1^2+t_2^2)(t_2-t_1) - 2(1+5t_1t_2)(t_2-t_1)$$
$$= 2(t_2-t_1)(4t_1^2+t_2^2-5t_1t_2), \quad (t_1,t_2) \in \mathbb{R}^2.$$

Therefore

$$\mathcal{G}(t_1,t_2) = \begin{pmatrix} 1+4t_1^2+t_2^2 & 1+5t_1t_2 \\ 1+5t_1t_2 & 1+t_1^2+4t_2^2 \end{pmatrix},$$

$$\mathcal{H}(t_1,t_2) = \frac{2(t_2-t_1)}{|t_1-t_2|(4(t_1+t_2)^2+5)^{1/2}} \begin{pmatrix} 1 & 1 \\ 1 & 1 \end{pmatrix},$$

$$\mathcal{A}(t_1,t_2) = \frac{2(t_2-t_1)}{|t_1-t_2|^3(4(t_1+t_2)^2+5)^{3/2}}$$
$$\times \begin{pmatrix} t_1^2+4t_2^2-5t_1t_2 & t_1^2+4t_2^2-5t_1t_2 \\ 4t_1^2+t_2^2-5t_1t_2 & 4t_1^2+t_2^2-5t_1t_2 \end{pmatrix}, \quad (t_1,t_2) \in \mathbb{R}^2.$$

Exercise 3.22. For the surface in Example 3.17 find
1. $\mathcal{G}(t_1,t_2)$, $(t_1,t_2) \in \mathbb{R}^2$;
2. $\mathcal{H}(t_1,t_2)$, $(t_1,t_2) \in \mathbb{R}^2$;
3. $\mathcal{A}(t_1,t_2)$, $(t_1,t_2) \in \mathbb{R}^2$.

Answer 3.8. 1.

$$\mathcal{G}(t_1,t_2) = \begin{pmatrix} 1+4t_1^2+9t_1^4 & 1+4t_1t_2+9t_1^2t_2^2 \\ 1+4t_1t_2+9t_1^2t_2^2 & 1+4t_2^2+9t_2^4 \end{pmatrix}, \quad (t_1,t_2) \in \mathbb{R}^2;$$

2.

$$\mathcal{H}(t_1,t_2) = \frac{1}{|t_2-t_1|(36t_1^2t_2+9(t_1+t_2)^2+4)^{1/2}}$$
$$\times \begin{pmatrix} 6(t_1-t_2)(t_2-11t_1) & 0 \\ 0 & 6(t_1-t_2)(t_2-11t_1) \end{pmatrix}, \quad (t_1,t_2) \in \mathbb{R}^2;$$

3.

$$A(t_1, t_2) = \frac{6(t_1 - t_2)(t_2 - 11t_1)}{|t_2 - t_1|^3 (36t_1^2 t_2 + 9(t_1 + t_2)^2 + 4)^{3/2}}$$
$$\times \begin{pmatrix} 1 + 4t_2^2 + 9t_2^4 & 1 + 4t_1 t_2 + 9t_1^2 t_2^2 \\ 1 + 4t_1 t_2 + 9t_1^2 t_2^2 & 1 + 4t_1 t_2 + 9t_1^2 t_2^2 \end{pmatrix}, \quad (t_1, t_2) \in \mathbb{R}^2.$$

3.11 The second fundamental form of a surface

Let S be an oriented surface.

Definition 3.29. For any $p \in S$, the second fundamental form is defined by

$$\phi_2(v, w) = -\phi_1(A(v), w), \quad v, w \in T_p S.$$

Exercise 3.23. For each $p \in S$, prove that the second fundamental form is a symmetric bilinear form.

Solution. Let $v, w \in T_p S$. Then

$$\begin{aligned}
\phi_2(v, w) &= -\phi_1(A(v), w) \\
&= -\phi_1(v, A(w)) \\
&= -\phi_1(A(w), v) \\
&= \phi_2(w, v).
\end{aligned}$$

Let now $v, \tilde{v}, w, \tilde{w} \in T_p S, \alpha_1, \alpha_2, \beta_1, \beta_2 \in \mathbb{R}$. Then

$$\begin{aligned}
\phi_2(\alpha_1 v + \alpha_2 \tilde{v}, w) &= -\phi_1(A(\alpha_1 v + \alpha_2 \tilde{v}), w) \\
&= -\phi_1(\alpha_1 A(v) + \alpha_2 A(\tilde{v}), w) \\
&= -\alpha_1 \phi_1(A(v), w) - \alpha_2 \phi_1(A(\tilde{v}), w) \\
&= \alpha_1 \phi_2(v, w) + \alpha_2 \phi_2(\tilde{v}, w)
\end{aligned}$$

and

$$\begin{aligned}
\phi_2(v, \beta_1 w + \beta_2 \tilde{w}) &= -\phi_1(A(v), \beta_1 w + \beta_2 \tilde{w}) \\
&= -\beta_1 \phi_1(A(v), w) - \beta_2 \phi_1(A(v), \tilde{w}) \\
&= \beta_1 \phi_2(v, w) + \beta_2 \phi_2(v, \tilde{w}).
\end{aligned}$$

This completes the solution.

Let (U, f) be a local representation of S that is compatible with the orientation of S. Then

$$\phi_2(f_{t_1}, f_{t_1}) = -\phi_1(A(f_{t_1}), f_{t_1})$$
$$= -\phi_1(n_{t_1}, f_{t_1})$$
$$= -n_{t_1} \cdot f_{t_1}$$
$$= L.$$

As above,

$$\phi_2(f_{t_1}, f_{t_2}) = -n_{t_1} \cdot f_{t_2} = M,$$
$$\phi_2(f_{t_2}, f_{t_1}) = -n_{t_2} \cdot f_{t_1} = M,$$
$$\phi_2(f_{t_2}, f_{t_2}) = -n_{t_2} \cdot f_{t_2} = N.$$

Consequently,

$$[\phi_2] = \begin{pmatrix} \phi_2(f_{t_1}, f_{t_1}) & \phi_2(f_{t_1}, f_{t_2}) \\ \phi_2(f_{t_1}, f_{t_2}) & \phi_2(f_{t_2}, f_{t_2}) \end{pmatrix}$$
$$= -\begin{pmatrix} n_{t_1} \cdot f_{t_1} & n_{t_1} \cdot f_{t_2} \\ n_{t_1} \cdot f_{t_2} & n_{t_2} \cdot f_{t_2} \end{pmatrix}$$
$$= \begin{pmatrix} L & M \\ M & N \end{pmatrix}$$

is the matrix of the second fundamental form ϕ_2. Recall here that

$$L = n \cdot f_{t_1 t_1},$$
$$M = n \cdot f_{t_1 t_2},$$
$$N = n \cdot f_{t_2 t_2}.$$

Example 3.27. Consider the surface in Example 3.26. Then the matrix of its second fundamental form is

$$[\phi_2](t_1, t_2) = \frac{2(t_2 - t_1)}{|t_1 - t_2|(4(t_1 + t_2)^2 + 5)^{1/2}} \begin{pmatrix} 1 & 1 \\ 1 & 1 \end{pmatrix}, \quad (t_1, t_2) \in \mathbb{R}^2.$$

Exercise 3.24. Find the matrix of the second fundamental form of the following surfaces:

1.
$$f(t_1, t_2) = (R \cos t_1 \cos t_2, R \cos t_1 \sin t_2, R \sin t_1), \quad (t_1, t_2) \in [0, 2\pi] \times [0, 2\pi],$$

where $R > 0$;

2.
$$f(t_1, t_2) = (a \cos t_1 \cos t_2, a \cos t_1 \sin t_2, c \sin t_2), \quad (t_1, t_2) \in [0, 2\pi] \times [0, 2\pi],$$

where $a, c \in \mathbb{R}$;

3.

$$f(t_1, t_2) = (a \cosh t_1 \cos t_2, a \cosh t_1 \sin t_2, c \sinh t_1), \quad t_1 \in \mathbb{R}, \quad t_2 \in [0, 2\pi],$$

where $a, c \in \mathbb{R}$;

4.

$$f(t_1, t_2) = (a \sinh t_1 \cos t_2, a \sinh t_1 \sin t_2, c \cosh t_1), \quad t_1 \in \mathbb{R}, \quad t_2 \in [0, 2\pi],$$

where $a, c \in \mathbb{R}$;

5.

$$f(t_1, t_2) = (t_1 \cos t_2, t_1 \sin t_2, t_1^2), \quad t_1 \in \mathbb{R}, \quad t_2 \in [0, 2\pi].$$

Answer 3.9. 1.

$$[\phi_2](t_1, t_2) = \frac{R}{|\cos t_1|} \begin{pmatrix} \cos t_1 & 0 \\ 0 & (\cos t_1)^3 \end{pmatrix}, \quad (t_1, t_2) \in [0, 2\pi] \times [0, 2\pi];$$

2.

$$[\phi_2](t_1, t_2) = \frac{a^2 c \sin t_1}{|a \sin t_1| \sqrt{a^2 (\cos t_1)^2 + c^2 (\cos t_2)^2}}$$

$$\times \begin{pmatrix} 0 & -\sin t_1 \cos t_2 \\ -\sin t_1 \cos t_2 & -\cos t_1 \sin t_2 \end{pmatrix}, \quad (t_1, t_2) \in [0, 2\pi] \times [0, 2\pi];$$

3.

$$[\phi_2](t_1, t_2) = \frac{a^2 c}{|a| \sqrt{a^2 (\sinh t_1)^2 + c^2 (\cosh t_1)^2}} \begin{pmatrix} -\cosh 2t_1 & 0 \\ 0 & (\cosh t_1)^2 \end{pmatrix}, \quad (t_1, t_2) \in \mathbb{R}^2;$$

4.

$$[\phi_2](t_1, t_2) = \frac{a^2 c \sinh t_1}{|a \sinh t_1| \sqrt{a^2 (\cosh t_1)^2 + c^2 (\sinh t_1)^2}}$$

$$\times \begin{pmatrix} 1 & 0 \\ 0 & (\sinh t_1)^2 \end{pmatrix}, \quad t_1 \in \mathbb{R}, \quad t_2 \in [0, 2\pi];$$

5.

$$[\phi_2](t_1, t_2) = \frac{2t_1}{|t_1| \sqrt{1 + 4t_1^2}} \begin{pmatrix} 1 & 0 \\ 0 & 2t_1^2 \end{pmatrix}, \quad t_1 \in \mathbb{R}, \quad t_2 \in [0, 2\pi].$$

3.12 The normal curvature. Meusnier theorem

Let S be a surface and (U, f) be a local parameterization that is compatible with the orientation of S. Suppose that $(I, g = g(t))$ be a parameterized curve lying on S and $n(g(t))$, $t \in I$, be the normal vector of S at $g(t)$. Set

$$\theta(t) = \angle(\mathbf{n}(t), n(g(t))), \quad t \in I,$$

where $\mathbf{n}$ is the principal normal vector field of g.

Definition 3.30. The normal curvature of g is defined as follows:

$$\kappa_n(t) = \kappa(t) \cos \theta(t), \quad t \in I,$$

where $\kappa(t)$ is the curvature of g.

Example 3.28. Consider a plane curve. Then $\theta(t) = \frac{\pi}{2}$ and $\cos \theta(t) = 0$. Hence, $\kappa_n(t) = 0$.

Example 3.29. Suppose that the support of a parameterized curve is a straight line. Then its normal curvature is zero, independently of the surface which includes the curve.

Exercise 3.25 (Meusnier theorem). Prove that

$$\kappa_n(t) = \phi_2(g'(t), g'(t))/\phi_1(g'(t), g'(t)), \quad t \in I.$$

Solution. Let g_1 be the naturally parameterized curve that is equivalent to g with arc length parameter s. Let also

$$g_1(s) = f(u(s), v(s)).$$

Then

$$g_1' = f_u u' + f_v v',$$
$$g_1'' = f_{uu}(u')^2 + f_{uv} u' v' + f_{uv} u' v' + f_{vv}(v')^2$$
$$= f_{uu}(u')^2 + 2f_{uv} u' v' + f_{vv}(v')^2.$$

Hence,

$$\kappa_n(s) = g_1''(s) \cdot n(g_1(s))$$
$$= (u')^2 (f_{uu} \cdot n) + 2u' v' (f_{uv} \cdot n) + (v')^2 (f_{vv} \cdot n)$$
$$= L(u')^2 - 2Mu' v' - N(v')^2$$
$$= \phi_2(g_1'(s), g_1'(s)).$$

Now, using that

$$g_1'(s) = \frac{g'(t)}{|g'(t)|},$$

we get

$$\kappa_n(t) = \phi_2\left(\frac{g'(t)}{|g'(t)|}, \frac{g'(t)}{|g'(t)|}\right)$$
$$= \phi_2\left(g'(t), \frac{g'(t)}{g'(t) \cdot g'(t)}\right)$$

$$= \frac{\phi_2(g'(t), g'(t))}{\phi_1(g'(t), g'(t))}, \quad t \in I.$$

This completes proof.

Exercise 3.26. If two curves on S have a common point and they have the same tangent lines at the common point, prove that they have the same normal curvature.

Solution. Let v and w be the tangent vectors to both curves. Then

$$w = \alpha v, \quad \alpha \in \mathbb{R}.$$

Hence,

$$\frac{\phi_2(w, w)}{\phi_1(w, w)} = \frac{\phi_2(\alpha v, \alpha v)}{\phi_1(\alpha v, \alpha v)}$$
$$= \frac{\alpha^2 \phi_2(v, v)}{\alpha^2 \phi_1(v, v)}$$
$$= \frac{\phi_2(v, v)}{\phi_1(v, v)}.$$

This completes the proof.

3.13 Asymptotic directions and lines

Let S be an oriented surface and (U, f) be its local representation that is compatible with the orientation of S.

Definition 3.31. For any $p \in S$, a direction $v \in T_pS$ is said to be asymptotic if

$$\phi_2(v, v) = 0.$$

Exercise 3.27. For any $p \in S$, a direction $v \in T_aS$ is asymptotic direction if and only if

$$LN - M^2 \le 0. \tag{3.7}$$

Solution. Let $v \ne 0$ and $v = (v_1, v_2) \in T_pS$. Without loss of generality, suppose that $v_2 \ne 0$. We have

$$0 = \phi_2(v, v)$$
$$= (v_1 v_2) \begin{pmatrix} L & M \\ M & N \end{pmatrix} \begin{pmatrix} v_1 \\ v_2 \end{pmatrix}$$
$$= (v_1 v_2) \begin{pmatrix} Lv_1 + Mv_2 \\ Mv_1 + Nv_2 \end{pmatrix}$$
$$= Lv_1^2 + 2Mv_1v_2 + Nv_2^2$$

if and only if

$$L(v_1/v_2)^2 + 2M(v_1/v_2) + N = 0$$

if and only if (3.7) holds. This completes the solution.

Definition 3.32. A point $p \in S$ is said to be
1.   elliptic if the second fundamental form is positive definite at p.
2.   hyperbolic if the second fundamental form is negative definite at p.
3.   parabolic if the determinant of the second fundamental form is 0 and at least one of its coefficients is different from 0 at p.
4.   planar if the coefficients of the second fundamental form are 0 at p.

3.14 Principal directions and curvatures. Gauss and mean curvatures

Suppose that S is an oriented surface and (U, f) is a local parameterization of S that is compatible with the orientation of S.

Definition 3.33. The directions on the tangent plane to S at a point $p \in S$ that are eigenvectors of the shape operator of S are called the principal directions of S.

Definition 3.34. A curve on S is called a principal line if its tangent directions at each point are principal directions.

Definition 3.35. A principal curvature of S is the normal curvature of S in a principal direction.

Exercise 3.28. Prove that the normal curvatures of S are the eigenvalues of the shape operator with opposite sign.

Solution. Let e be an eigenvector of A. Then there is a constant $\lambda \in \mathbb{R}$ such that

$$A(e) = \lambda e.$$

Hence,

$$
\begin{aligned}
\kappa_n(e) &= \frac{\phi_2(e, e)}{\phi_1(e, e)} \\
&= \frac{A(e) \cdot e}{e \cdot e} \\
&= \lambda \frac{e \cdot e}{e \cdot e} \\
&= \lambda.
\end{aligned}
$$

This completes the proof.

In what follows, suppose that k_1 and k_2 are the principal curvatures of S and $k_1 \geq k_2$. Let $\{e_1, e_2\}$ be an orthonormal basis of the principal directions. Then

$$A(e_j) = -k_j e_j, \quad j \in \{1, 2\}.$$

Let $|e| = 1$. Then

$$e = e_1 \cos \theta + e_2 \sin \theta,$$

and

$$\begin{aligned} \phi_2(e, e) &= -(A(e_1 \cos \theta + e_2 \sin \theta) \cdot (e_1 \cos \theta + e_2 \sin \theta)) \\ &= -((-k_1 e_1 \cos \theta - k_2 e_2 \sin \theta) \cdot (e_1 \cos \theta + e_2 \sin \theta)) \\ &= k_1 (\cos \theta)^2 + k_2 (\sin \theta)^2, \end{aligned}$$

as well as

$$\phi_1(e, e) = 1.$$

Hence,

$$\kappa_n(e) = k_1 (\cos \theta)^2 + k_2 (\sin \theta)^2. \tag{3.8}$$

Definition 3.36. Equation (3.8) is called the Euler formula.

Definition 3.37. The quantity

$$K = k_1 k_2$$

is called the Gaussian curvature of S.

Definition 3.38. The quantity

$$H = \frac{k_1 + k_2}{2}$$

is called the mean curvature of S.

We have that

$$K = \det \mathcal{A}$$

and

$$H = \frac{1}{2} \operatorname{Tr} \mathcal{A}.$$

Exercise 3.29 (Joachimsthal theorem). Let S_1 and S_2 be two oriented surfaces and y be a parameterized curve that lies on the intersection of S_1 and S_2. Let also S_1 and S_2 intersect under a constant angle. Prove that y is a curvature line on S_1 if and only if y_2 is a curvature line on S_2.

Solution. Let $(I, f = f(t))$ be a local parameterization of y. Let also n_1 and n_2 be normal directions to S_1 and S_2, respectively. Then

$$n_1 \cdot n_2 = \text{const.}$$

Hence,

$$0 = n_1' \cdot n_2 + n_1 \cdot n_2'.$$

Suppose that y is a curvature line on S_1. Then

$$n_1' = -lf',$$

where l is one of the principal curvatures of S_1. Since y lies on S_2, we have that

$$f' \cdot n_2 = 0.$$

Thus,

$$0 = (-lf') \cdot n_2 + n_1 \cdot n_2'$$
$$= n_1 \cdot n_2'.$$

Hence,

$$n_1 \cdot n_2' = 0.$$

Since $n_2 \perp n_2'$ and $f' \perp n_2$, we conclude that $n_2 \parallel f'$ and there is an $m \in \mathbb{R}$ such that

$$n_2 = -mf',$$

i. e., y is a curvature line on S_2. This completes the proof.

Let S be an oriented surface and (U, f) be a local representation of S that is compatible with the orientation of S.

Exercise 3.30. Prove that

$$K = \frac{LN - M^2}{W^2},$$
$$H = \frac{GL - 2FM + EN}{2W^2}.$$

Solution. We have

$$A = \frac{1}{W^2} \begin{pmatrix} GL - FM & GM - FN \\ EM - FL & EN - FM \end{pmatrix}.$$

Then

$$\det A = \frac{1}{W^4}(GELN - GFLM - FEMN + F^2M^2 - EGM^2$$
$$+ EFMN - F^2LN + FGLM)$$
$$= \frac{1}{W^4}(EG(LN - M^2) - (LN - M^2)F^2)$$
$$= \frac{1}{W^4}((EG - F^2)(LN - M^2))$$
$$= \frac{1}{W^2}(LN - M^2).$$

Next,

$$H = \frac{1}{2}\operatorname{Tr} A$$
$$= \frac{GL - 2FM + EN}{2W^2}.$$

This completes the solution.

3.15 Advanced practical problems

Problem 3.1. Prove that the hyperbolic paraboloid

$$f(t_1, t_2) = (a(t_1 + t_2), b(t_2 - t_1), 2t_1t_2), \quad (t_1, t_2) \in \mathbb{R}^2,$$

where $a, b \in \mathbb{R}$, $a > 0$, $b > 0$, is a regular surface. See Fig. 3.19 for the values of $a = b = 1$.

Problem 3.2. Find the support of the surface

$$f(t_1, t_2) = \left(\frac{t_1}{t_1^2 + t_2^2 + 1}, \frac{t_2}{t_1^2 + t_2^2 + 1}, \frac{t_1^2 + t_2^2}{t_1^2 + t_2^2 + 1} \right), \quad (t_1, t_2) \in \mathbb{R}^2.$$

Answer 3.10. The sphere

$$f_1^2 + f_2^2 + f_3^2 = f_3$$

or

$$f_1^2 + f_2^2 + \left(f_3 - \frac{1}{2} \right)^2 = \frac{1}{4}.$$

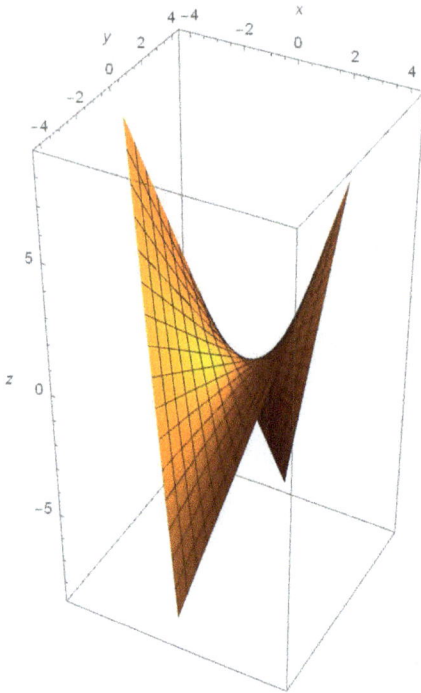

Figure 3.19: The surface $f(t_1, t_2) = (t_1 + t_2, t_2 - t_1, 2t_1t_2)$, $(t_1, t_2) \in [-2, 2]$.

Problem 3.3. Find the implicit equation of the surface

$$(x_0 + a \cos t_1 \cos t_2, y_0 + b \sin t_1 \cos t_2, z_0 + c \sin t_2), \quad t_1 \in [0, 2\pi], \quad t_2 \in [0, \pi],$$

where $x_0, y_0, z_0, a, b, c \in \mathbb{R}$, $(a, b, c) \neq (0, 0, 0)$.

Answer 3.11.

$$\frac{(x - x_0)^2}{a^2} + \frac{(y - y_0)^2}{b^2} + \frac{(z - z_0)^2}{c^2} = 1.$$

Problem 3.4. Find the equation of the tangent plane to the following surfaces at the given points:
1.

$$z = \sqrt{x^2 + y^4}, \quad (0, 0, 0);$$

2.

$$z = x - y + \sqrt{|xy|}, \quad (0, 0, 0);$$

3.

$$z = \log \sqrt{x^2 + y^2}, \quad (0, 1, 0);$$

4.
$$z = \sin\left(\frac{x}{y}\right), \quad (\pi, 1, 0);$$

5.
$$z = e^{x \cos y}, \quad (1, 0, e);$$

6.
$$z = \arctan\left(\frac{y}{x}\right), \quad \left(1, 1, \frac{\pi}{4}\right);$$

7.
$$z = y + \log\left(\frac{x}{z}\right), \quad (1, 1, 1);$$

8.
$$2^{\frac{x}{z}} + 2^{\frac{y}{z}} = 8, \quad (2, 2, 1);$$

9.
$$(t_1 + \log t_2, t_2 - \log t_1, 2t_1 + t_2), \quad (1, 1, 3);$$

10.
$$(\cos t_1 \cosh t_2, \sin t_1 \cosh t_2, \sinh t_2), \quad \left(\cosh\left(\frac{1}{\sqrt{2}}\right), 0, \sinh\left(\frac{1}{\sqrt{2}}\right)\right);$$

11.
$$(t_1 + t_2, t_1 - t_2, t_1 t_2), \quad (2, 1);$$

12.
$$(t_1, t_1^2 - 2t_2, t_1^3 - 3t_1 t_2), \quad (1, 3, 4);$$

13.
$$(t_1 \cos t_2, t_1 \sin t_2, t_1), \quad \left(2, \frac{\pi}{4}\right);$$

14.
$$\frac{x^2}{a^2} + \frac{y^2}{b^2} + \frac{z^2}{c^2} = 1, \quad (x_0, y_0, z_0);$$

15.
$$(t_1^0 \cos t_2^0, t_1^0 \sin t_2^0, a t_2^0), \quad (t_1^0, t_2^0).$$

Answer 3.12. 1. Not exist;
2.   Not exist;
3.
$$y - z = 1;$$

4.
$$x - \pi y + z = 0;$$

5.
$$ex - z = 0;$$

6.
$$x - y + 2z = \frac{\pi}{2};$$

7.
$$x + y - 2z = 0;$$

8.
$$x + y - 4z = 0;$$

9.
$$3x - y - 2z + 4 = 0;$$

10.
$$\cosh\left(\frac{1}{\sqrt{2}}\right)x - \sinh\left(\frac{1}{\sqrt{2}}\right)z = 1;$$

11.
$$3x - y - 2z = 4;$$

12.
$$6x + 3y - 2z = 7;$$

13.
$$x + y - \sqrt{2}z = 0;$$

14.
$$\frac{xx_0}{a^2} + \frac{yy_0}{b^2} + \frac{zz_0}{c^2} = 1;$$

15.
$$(a\sin t_2^0)x - a(\cos t_2^0)y + (t_2^0)z - at_1^0 t_2^0 = 0.$$

Problem 3.5. Find the equation of the normal line to the surfaces in Problem 3.4 at the given points.

Answer 3.13. 1. Does not exist;
2. Does not exist;
3.
$$\frac{x}{0} = y - 1 = -z;$$

4.
$$x - \pi = \frac{y-1}{-\pi} = z;$$

5.
$$\frac{x-1}{e} = \frac{y}{0} = \frac{z-e}{-1};$$

6.
$$x - 1 = \frac{y-1}{-1} = \frac{z - \frac{\pi}{4}}{2};$$

7.
$$x - 1 = y - 1 = \frac{z-1}{-2};$$

8.
$$x - 2 = y - 2 = \frac{z-1}{-4};$$

9.
$$\frac{x-1}{-3} = y - 1 = \frac{z-3}{2};$$

10.
$$\frac{x - \cosh(\frac{1}{\sqrt{2}})}{\cosh^2(\frac{1}{\sqrt{2}})} = \frac{y}{0} = \frac{-z + \sinh(\frac{1}{\sqrt{2}})}{\cosh(\frac{1}{\sqrt{2}})\sinh(\frac{1}{\sqrt{2}})};$$

11.
$$\frac{x-2}{3} = -y+1 = \frac{z-2}{-2};$$

12.
$$\frac{x-1}{-18} = \frac{y-3}{3} = \frac{z-4}{2};$$

13.
$$\frac{x-\sqrt{2}}{-\sqrt{2}} = \frac{y-\sqrt{2}}{-\sqrt{2}} = \frac{z-2}{2};$$

14.
$$\frac{x-x_0}{\frac{2x_0}{a^2}} = \frac{y-y_0}{\frac{2y_0}{b^2}} = \frac{z-z_0}{\frac{2z_0}{c^2}};$$

15.
$$\frac{x-t_1^0\cos t_2^0}{a\sin t_2^0} = \frac{y-t_1^0\sin t_2^0}{-a\cos t_2^0} = \frac{z-at_2^0}{t_1^0}.$$

Problem 3.6. Find the matrix of the first fundamental form for the following surfaces:

1.
$$(t_1\cos t_2, t_1\sin t_2, kt_1), \quad t_1 \in \mathbb{R}, \quad v \in [0, 2\pi],$$

where $k \in \mathbb{R}$;

2.
$$((a+b\cos t_1)\cos t_2, (a+b\cos t_1)\sin t_2, b\sin t_1), \quad t_1, t_2 \in [0, 2\pi],$$

where $a, b \in \mathbb{R}$;

3.
$$\left(a\cosh\frac{t_1}{a}\cos t_2, a\cosh\frac{t_1}{a}\sin t_2, t_1\right), \quad t_1 \in \mathbb{R}, \quad t_2 \in [0, 2\pi],$$

where $a \in \mathbb{R}, a \neq 0$;

4.
$$\left(a\sin t_1\cos t_2, a\sin t_1\sin t_2, a\left(\log\left(\tan\frac{t_1}{2}\right) + \cos t_1\right)\right),$$

$t_1, t_2 \in [0, 2\pi]$, where $a \in \mathbb{R}$;

5.
$$(t_1\cos t_2, t_1\sin t_2, at_2), \quad t_1 \in \mathbb{R}, \quad v \in [0, 2\pi],$$

where $a \in \mathbb{R}$;

6.
$$(t_1\cos t_2, t_1\sin t_2, h(t_1) + at_2), \quad (t_1, t_2) \in \mathbb{R}^2,$$

where $h \in C^1(\mathbb{R})$ and $a \in \mathbb{R}$.

Answer 3.14. 1.
$$\begin{pmatrix} 1+k^2 & 0 \\ 0 & t_1^2 \end{pmatrix}, \quad t_1 \in \mathbb{R}, \quad t_2 \in [0, 2\pi];$$

2.
$$\begin{pmatrix} b^2 & 0 \\ 0 & (a + b\cos t_1)^2 \end{pmatrix}, \quad t_1, t_2 \in [0, 2\pi];$$

3.
$$\begin{pmatrix} (\cosh \frac{t_1}{a})^2 & 0 \\ 0 & (a\cosh \frac{t_1}{a})^2 \end{pmatrix}, \quad t_1 \in \mathbb{R}, \quad t_2 \in [0, 2\pi];$$

4.
$$\begin{pmatrix} a^2(\cot t_1)^2 & 0 \\ 0 & a^2(\sin t_1)^2 \end{pmatrix}, \quad t_1, t_2 \in [0, 2\pi];$$

5.
$$\begin{pmatrix} 1 & 0 \\ 0 & a^2 + t_1^2 \end{pmatrix}, \quad t_1 \in \mathbb{R}, \quad t_2 \in [0, 2\pi];$$

6.
$$\begin{pmatrix} 1 + (h'(t_1))^2 & ah'(t_1) \\ ah'(t_1) & a^2 + t_1^2 \end{pmatrix}, \quad (t_1, t_2) \in \mathbb{R}^2.$$

Problem 3.7. Let S be a surface and the matrix of its first fundamental form be

$$\begin{pmatrix} 1 & 0 \\ 0 & (\sinh t_1)^2 \end{pmatrix}, \quad (t_1, t_2) \in \mathbb{R}^2.$$

Find the length of the segment of the curve

$$t_1 = t_2$$

between the points (t_1^0, t_2^0) and (t_1^1, t_2^1).

Answer 3.15.

$$|\sinh t_1^1 - \sinh t_1^0|.$$

Problem 3.8. On the helicoid

$$(t_1 \cos t_2, t_1 \sin t_2, at_2), \quad t_1 \in \mathbb{R}, \quad t_2 \in [0, 2\pi],$$

where $a \in \mathbb{R}$, consider the following curves:

$$t_2 = \log(t_1 \pm \sqrt{t_1^2 + a^2}) + c,$$

where $a, c \in \mathbb{R}$. Find the lengths of the segments of the curve between the points (t_1^0, t_2^0) and (t_1^1, t_2^1).

Answer 3.16.

$$\sqrt{2}|t_1^1 - t_1^0|.$$

Problem 3.9. On the pseudosphere

$$\left(a \sin t_1 \cos t_2, a \sin t_1 \sin t_2, a\left(\log\left(\tan \frac{t_1}{2} \right) + \cos t_1 \right) \right), \quad t_1, t_2 \in [0, 2\pi],$$

where $a \in \mathbb{R}$, consider the curves

$$t_2 = \pm a \log\left(\tan \frac{t_1}{2} \right) + c,$$

where $c \in \mathbb{R}$. Find the lengths of the segments of the curves between the points (t_1^0, t_2^0) and (t_1^1, t_2^1).

Answer 3.17.

$$\frac{|c_2 - c_1|}{2},$$

where

$$t_2^0 = \frac{c_1 + c}{2},$$
$$t_2^1 = \frac{c_2 + c}{2}.$$

Problem 3.10. Find the angle between the curves

$$t_2 = t_1 + 1 \quad \text{and} \quad t_2 = 3 - t_1$$

on the surface

$$f(t_1, t_2) = (t_1 \cos t_2, t_1 \sin t_2, t_1^2), \quad t_1 \in \mathbb{R}, \quad t_2 \in [0, 2\pi].$$

Answer 3.18.

$$\cos \theta = \frac{2}{3}.$$

Problem 3.11. Find the area of the curvilinear triangle

$$t_1 = \pm a t_2, \quad t_2 = 1$$

on the surface with a matrix of its first fundamental form given by

$$\begin{pmatrix} 1 & 0 \\ 0 & t_1^2 + a^2 \end{pmatrix}, \quad (t_1, t_2) \in \mathbb{R}^2.$$

Here $a \in \mathbb{R}$ is a given parameter.

Answer 3.19.

$$a^2 \left(\frac{2}{3} - \frac{\sqrt{2}}{3} + \log(1 + \sqrt{2}) \right).$$

Problem 3.12. Consider the helicoid

$$f(t_1, t_2) = (t_1 \cos t_2, t_1 \sin t_2, at_1), \quad (t_1, t_2) \in \mathbb{R}^2,$$

where $a > 0$. Find:
1. $\mathcal{G}(t_1, t_2), (t_1, t_2) \in \mathbb{R}^2$;
2. $\mathcal{H}(t_1, t_2), (t_1, t_2) \in \mathbb{R}^2$;
3. $\mathcal{A}(t_1, t_2), (t_1, t_2) \in \mathbb{R}^2$.

Answer 3.20. 1.

$$\mathcal{G}(t_1, t_2) = \begin{pmatrix} 1 & 0 \\ 0 & t_1^2 + a^2 \end{pmatrix}, \quad (t_1, t_2) \in \mathbb{R}^2;$$

2.

$$\mathcal{H}(t_1, t_2) = \begin{pmatrix} 0 & 0 \\ 0 & \frac{\operatorname{sign}(u)}{\sqrt{a^2 + u^2}}(au) \end{pmatrix}, \quad (t_1, t_2) \in \mathbb{R}^2;$$

3.

$$\mathcal{A}(t_1, t_2) = \frac{-1}{u^2 + a^2} \begin{pmatrix} 0 & \frac{\operatorname{sign}(t_1)}{\sqrt{a^2 + t_1^2}}(at_1) \\ 0 & \frac{\operatorname{sign}(t_1)}{\sqrt{a^2 + t_1^2}}(at_1) \end{pmatrix}, \quad (t_1, t_2) \in \mathbb{R}^2.$$

Problem 3.13. Find the matrix of the second fundamental form of the following surfaces:
1.

$$f(t_1, t_2) = (R \cos t_2, R \sin t_2, t_1), \quad t_1, t_2 \in [0, 2\pi],$$

where $R > 0$;
2.

$$f(t_1, t_2) = (t_1 \cos t_2, t_1 \sin t_2, kt_1), \quad t_1, t_2 \in [0, 2\pi];$$

3.

$$f(t_1, t_2) = ((a + b \cos t_1) \cos t_2, (a + b \cos t_1) \sin t_2, b \sin t_1), \quad t_1, t_2 \in [0, 2\pi],$$

where $a, b \in \mathbb{R}, (a, b) \neq (0, 0)$;
4.

$$f(t_1, t_2) = \left(a \cosh \frac{t_1}{a} \cos t_2, a \cosh \frac{t_1}{a} \sin t_2, t_1 \right), \quad t_1 \in \mathbb{R}, \quad t_2 \in [0, 2\pi];$$

5.

$$f(t_1, t_2) = \left(a \sin t_1 \cos t_2, a \sin t_1 \sin t_2, a \left(\log \left(\tan \frac{t_1}{2} \right) + \cos t_1 \right) \right), \quad t_1, t_2 \in [0, 2\pi].$$

Answer 3.21. 1.

$$[\phi_2](t_1, t_2) = \begin{pmatrix} 0 & 0 \\ 0 & R \end{pmatrix}, \quad t_1 \in \mathbb{R}, \quad t_2 \in [0, 2\pi];$$

2.
$$[\phi_2](t_1, t_2) = \begin{pmatrix} 0 & 0 \\ 0 & \frac{\text{sign}(t_1)}{\sqrt{1+k^2}}(ku) \end{pmatrix}, \quad t_1 \in \mathbb{R}, \quad t_2 \in [0, 2\pi];$$

3.
$$[\phi_2](t_1, t_2) = \text{sign}(b(a + b \cos t_1)) \begin{pmatrix} b & 0 \\ 0 & \cos t_1(a + b \cos t_1) \end{pmatrix}, \quad t_1, t_2 \in [0, 2\pi];$$

4.
$$[\phi_2](t_1, t_2) = \text{sign}(a) \cosh \frac{t_1}{a} \begin{pmatrix} \frac{1}{a} & 0 \\ 0 & a \end{pmatrix}, \quad t_1 \in \mathbb{R}, \quad t_2 \in [0, 2\pi];$$

5.
$$[\phi_2](t_1, t_2) = \begin{pmatrix} -a\text{sign}(a) \cot t_1 & 0 \\ 0 & a\text{sign}(a) \sin t_1 \cos t_1 \end{pmatrix}, \quad t_1, t_2 \in [0, 2\pi].$$

Problem 3.14. Find the principal curvatures of the following surfaces at the given points:

1.
$$z = xy, \quad (1, 1, 1);$$

2.
$$\frac{x^2}{p} + \frac{y^2}{q} = 2z, \quad (0, 0, 0);$$

3.
$$(t_1^2 + t_2^2, t_1^2 - t_2^2, t_1 t_2), \quad (1, 1).$$

Answer 3.22. 1.
$$k_1 = -\frac{\sqrt{3}}{9},$$
$$k_2 = \frac{\sqrt{3}}{3}.$$

2.
$$k_1 = 2p,$$
$$k_2 = 2q.$$

3.
$$k_1 = \frac{1}{2\sqrt{5}},$$
$$k_2 = 0.$$

Problem 3.15. Find the Gauss and mean curvatures of the following surfaces:

1.
$$z = f(x, y), \quad (x, y) \in \mathbb{R}^2;$$

2.
$$z = f(\sqrt{x^2 + y^2}), \quad (x, y) \in \mathbb{R}^2.$$

Answer 3.23. 1.

$$K = \frac{z_{xx}z_{yy} - (z_{xy})^2}{(1 + z_x^2 + z_y^2)^2},$$

$$H = \frac{(1 + z_x^2)z_{yy} + (1 + z_y^2)z_{xx} - 2z_xz_yz_{xy}}{(1 + z_x^2 + z_y^2)^{\frac{3}{2}}};$$

2.

$$K = \frac{f'f''}{\sqrt{x^2 + y^2}(1 + (f')^2)^3},$$

$$H = \frac{f''}{4(1 + (f')^2)^{\frac{3}{2}}} + \frac{f'}{4\sqrt{x^2 + y^2}\sqrt{1 + (f')^2}}.$$

Problem 3.16. Let S be a surface with a matrix of its first fundamental form being

$$\begin{pmatrix} 1 & \cos\omega \\ \cos\omega & 1 \end{pmatrix}.$$

Prove that

$$K = \frac{\omega_{t_1 t_2}}{\sin\omega}.$$

Problem 3.17. Let S be a surface with a matrix of its first fundamental form being

$$\begin{pmatrix} \frac{1}{(x^2+y^2+c^2)^2} & 0 \\ 0 & \frac{1}{(x^2+y^2+c^2)^2} \end{pmatrix}.$$

Prove that K is a constant.

Problem 3.18. Classify the points of the following surfaces:
1. ellipsoid;
2. elliptic paraboloid;
3. hyperbolic paraboloid;
4. elliptic cylinder;
5. parabolic cylinder;
6. hyperbolic cylinder;
7. cone;
8.

$$z = f(\sqrt{x^2 - y^2}), \quad (x,y) \in \mathbb{R}^2;$$

9.

$$x + y = z^3, \quad (x,y) \in \mathbb{R}^2.$$

Answer 3.24. 1. elliptic;
2. elliptic;

3. hyperbolic;
4. parabolic;
5. parabolic;
6. parabolic;
7. parabolic;
8. if $f'f'' < 0$, then elliptic; if $f'f'' > 0$, then hyperbolic; if $f'f'' = 0$, then parabolic;
9. parabolic.

4 Fundamental equations of a surface. Special classes of surfaces

In this chapter introducing the Christoffel symbols, the fundamental equations of a surface, Gauss and Codazzi–Mainardi equations, are established, which are important due to their role as existence and uniqueness theorems for surfaces. Geodesics playing the role of a line on a surface are introduced as a notable class of curves on surfaces. Some special surfaces such as Liouville, ruled, and minimal surfaces are also given.

4.1 Some relations

Suppose that (U, f) is a regular parameterized surface. Then $\{f_{t_1}, f_{t_2}, n\}$ is a basis in $\mathbb{R}^3_{f(t_1, t_2)}$. We have

$$E = f_{t_1} \cdot f_{t_1},$$
$$F = f_{t_1} \cdot f_{t_2},$$
$$G = f_{t_2} \cdot f_{t_2}.$$

Then

$$E_{t_1} = f_{t_1 t_1} \cdot f_{t_1} + f_{t_1} \cdot f_{t_1 t_1}$$
$$= 2(f_{t_1 t_1} \cdot f_{t_1}),$$

whereupon

$$f_{t_1 t_1} \cdot f_{t_1} = \frac{1}{2} E_{t_1}.$$

Also,

$$E_{t_2} = f_{uv} \cdot f_{t_1} + f_{t_1} \cdot f_{t_1 t_2}$$
$$= 2(f_{t_1} \cdot f_{t_1 t_2})$$

and

$$f_{t_1} \cdot f_{t_1 t_2} = \frac{1}{2} E_{t_2}.$$

Consider the component G. We have

$$G_{t_1} = f_{t_1 t_2} \cdot f_{t_2} + f_{t_2} \cdot f_{t_1 t_2}$$
$$= 2(f_{t_1 t_2} \cdot f_{t_2}),$$

https://doi.org/10.1515/9783111501857-004

whereupon

$$f_{t_1 t_2} \cdot f_{t_2} = \frac{1}{2} G_{t_1},$$

and

$$G_{t_2} = f_{t_2 t_2} \cdot f_{t_2} + f_{t_2} \cdot f_{t_2 t_2}$$
$$= 2(f_{t_2 t_2} \cdot f_{t_2}),$$

from where

$$f_{t_2 t_2} \cdot f_{t_2} = \frac{1}{2} G_{t_2}.$$

Moreover,

$$F_{t_1} = f_{t_1 t_1} \cdot f_{t_2} + f_{t_1} \cdot f_{t_1 t_2}$$
$$= f_{t_1 t_1} \cdot f_{t_2} + \frac{1}{2} E_{t_2},$$

whereupon

$$f_{t_1 t_1} \cdot f_{t_2} = F_{t_1} - \frac{1}{2} E_{t_2},$$

and

$$F_{t_2} = f_{t_1 t_2} \cdot f_{t_2} + f_{t_1} \cdot f_{t_2 t_2}$$
$$= \frac{1}{2} G_{t_1} + (f_{t_1} \cdot f_{t_2 t_2}),$$

from where

$$f_{t_1} \cdot f_{t_2 t_2} = F_{t_2} - \frac{1}{2} G_{t_1}.$$

4.2 The Christoffel symbols

In this section, we will represent $f_{t_1 t_1}, f_{t_1 t_2}$, and $f_{t_2 t_2}$ in terms of the basis $\{f_{t_1}, f_{t_2}, n\}$ in the following way:

$$f_{t_1 t_1} = \Gamma^1_{11} f_{t_1} + \Gamma^2_{11} f_{t_2} + Ln,$$
$$f_{t_1 t_2} = \Gamma^1_{12} f_{t_1} + \Gamma^2_{12} f_{t_2} + Mn, \tag{4.1}$$
$$f_{t_2 t_2} = \Gamma^1_{22} f_{t_1} + \Gamma^2_{22} f_{t_2} + Nn,$$

where L, M, N are the components of the second fundamental form.

Definition 4.1. The coefficients Γ_{ij}^k, $i, j, k \in \{1, 2\}$, are called the Christoffel symbols.

From the first equation of the system (4.1), we find

$$f_{t_1 t_1} \cdot f_{t_1} = \Gamma_{11}^1 (f_{t_1} \cdot f_{t_1}) + \Gamma_{11}^2 (f_{t_1} \cdot f_{t_2}),$$
$$f_{t_1 t_1} \cdot f_{t_2} = \Gamma_{11}^1 (f_{t_1} \cdot f_{t_2}) + \Gamma_{11}^2 (f_{t_2} \cdot f_{t_2}),$$

or

$$\Gamma_{11}^1 E + \Gamma_{11}^2 F = \frac{1}{2} E_{t_1},$$
$$\Gamma_{11}^1 F + \Gamma_{11}^2 G = F_{t_1} - \frac{1}{2} E_{t_2}.$$

Since $W^2 = EG - F^2 \neq 0$, the above system has a unique solution, namely

$$\Gamma_{11}^1 = \frac{\frac{1}{2} E_{t_1} G - F_{t_1} F + \frac{1}{2} E_{t_2} F}{W^2},$$
$$\Gamma_{11}^2 = \frac{-\frac{1}{2} E_{t_1} F + E F_{t_1} - \frac{1}{2} E E_{t_2}}{W^2}.$$

Analogously, using the second equation of the system (4.1), we obtain

$$f_{t_1 t_2} \cdot f_{t_1} = \Gamma_{12}^1 (f_{t_1} \cdot f_{t_1}) + \Gamma_{12}^2 (f_{t_1} \cdot f_{t_2}),$$
$$f_{t_1 t_2} \cdot f_{t_2} = \Gamma_{12}^1 (f_{t_1} \cdot f_{t_2}) + \Gamma_{12}^2 (f_{t_2} \cdot f_{t_2}),$$

or

$$\Gamma_{12}^1 E + \Gamma_{12}^2 F = \frac{1}{2} E_{t_2},$$
$$\Gamma_{12}^1 F + \Gamma_{12}^2 G = \frac{1}{2} G_{t_1}.$$

The solutions are

$$\Gamma_{12}^1 = \frac{E_{t_2} G - G_{t_1} F}{2W^2},$$
$$\Gamma_{12}^2 = \frac{-E_{t_2} F + E G_{t_1}}{2W^2}.$$

Finally, from the third equation of the system (4.1), we obtain

$$f_{t_2 t_2} \cdot f_{t_1} = \Gamma_{22}^1 (f_{t_1} \cdot f_{t_1}) + \Gamma_{22}^2 (f_{t_1} \cdot f_{t_2}),$$
$$f_{t_2 t_2} \cdot f_{t_2} = \Gamma_{22}^1 (f_{t_1} \cdot f_{t_2}) + \Gamma_{22}^2 (f_{t_2} \cdot f_{t_2}),$$

or

$$\Gamma^1_{22}E + \Gamma^2_{22}F = F_{t_2} - \frac{1}{2}G_{t_1},$$

$$\Gamma^1_{22}F + \Gamma^2_{22}G = \frac{1}{2}G_{t_2}.$$

The solutions are

$$\Gamma^1_{22} = \frac{F_{t_2}G - \frac{1}{2}G_{t_1}G - \frac{1}{2}G_{t_2}F}{W^2},$$

$$\Gamma^2_{22} = \frac{-FF_{t_2} + \frac{1}{2}FG_{t_1} + \frac{1}{2}EG_{t_2}}{W^2}.$$

Example 4.1. Let $f, g \in C^2(I)$, $(f'(t), g'(t)) \neq (0,0)$, $t \in I$, $I \subset \mathbb{R}$. Assume that S is a surface obtained by rotating the following curve:

$$x = u(t),$$
$$z = v(t), \quad t \in I,$$

around the z-axis. Then S can be represented parametrically as follows:

$$f(t, \theta) = (u(t)\cos\theta, u(t)\sin\theta, v(t)), \quad t \in I, \quad \theta \in [0, 2\pi].$$

We have

$$f_t(t, \theta) = (u'(t)\cos\theta, u'(t)\sin\theta, v'(t)),$$
$$f_\theta(t, \theta) = (-u(t)\sin\theta, u(t)\cos\theta, 0), \quad t \in I, \quad \theta \in [0, 2\pi],$$

and

$$
\begin{aligned}
E(t, \theta) &= f_t(t, \theta) \cdot f_t(t, \theta) \\
&= (u'(t)\cos\theta)^2 + (u'(t)\sin\theta)^2 + (v'(t))^2 \\
&= (u'(t))^2(\cos\theta)^2 + (u'(t))^2(\sin\theta)^2 + (v'(t))^2 \\
&= (u'(t))^2 + (v'(t))^2, \\
F(t, \theta) &= f_t(t, \theta) \cdot f_\theta(t, \theta) \\
&= -u(t)u'(t)\sin\theta\cos\theta + u(t)u'(t)\sin\theta\cos\theta \\
&= 0, \\
G(t, \theta) &= f_\theta(t, \theta) \cdot f_\theta(t, \theta) \\
&= (-u(t)\sin\theta)^2 + (u(t)\cos\theta)^2 \\
&= (u(t))^2(\cos\theta)^2 + (u(t))^2(\sin\theta)^2 \\
&= (u(t))^2, \quad t \in I, \quad \theta \in [0, 2\pi],
\end{aligned}
$$

and

$$E_t(t, \theta) = 2u'(t)u''(t) + 2v'(t)v''(t),$$
$$E_\theta(t, \theta) = 0,$$
$$F_t(t, \theta) = 0,$$
$$F_\theta(t, \theta) = 0,$$
$$G_t(t, \theta) = 2u(t)u'(t),$$
$$G_\theta(t, \theta) = 0, \quad t \in I, \quad \theta \in [0, 2\pi],$$

as well as

$$W(t, \theta) = E(t, \theta)G(t, \theta) - (F(t, \theta))^2$$
$$= (u(t))^2 ((u'(t))^2 + (v'(t))^2), \quad t \in I, \quad \theta \in [0, 2\pi].$$

Hence,

$$\Gamma_{11}^1(t, \theta) = \frac{\frac{1}{2}E_t(t, \theta)G(t, \theta) - F_t(t, \theta)F(t, \theta) + \frac{1}{2}E_\theta(t, \theta)F(t, \theta)}{(W(t, \theta))^2}$$

$$= \frac{\frac{1}{2}(u(t))^2 (2u'(t)u''(t) + 2v'(t)v''(t))}{(u(t))^2 ((u'(t))^2 + (v'(t))^2)}$$

$$= \frac{u'(t)u''(t) + v'(t)v''(t)}{(u'(t))^2 + (v'(t))^2},$$

$$\Gamma_{11}^2(t, \theta) = \frac{-\frac{1}{2}E_t(t, \theta)F(t, \theta) + E(t, \theta)F_t(t, \theta) - \frac{1}{2}E(t, \theta)E_\theta(t, \theta)}{(W(t, \theta))^2} = 0,$$

$$\Gamma_{12}^1(t, \theta) = \frac{E_\theta(t, \theta)G(t, \theta) - G_t(t, \theta)F(t, \theta)}{2(W(t, \theta))^2} = 0,$$

$$\Gamma_{12}^2(\theta, t) = \frac{-E_\theta(t, \theta)F(t, \theta) + E(t, \theta)G_t(t, \theta)}{2(W(t, \theta))^2}$$

$$= \frac{((u'(t))^2 + (v'(t))^2)2u(t)u'(t)}{(u(t))^2 ((u'(t))^2 + (v'(t))^2)},$$

$$= \frac{u'(t)}{u(t)},$$

$$\Gamma_{22}^1(t, \theta) = \frac{F_\theta(t, \theta)G(t, \theta) - \frac{1}{2}G_t(t, \theta)G(t, \theta) - \frac{1}{2}G_\theta(t, \theta)F(t, \theta)}{(W(t, \theta))^2}$$

$$= \frac{-\frac{1}{2}(u(t))^2 2u(t)u'(t)}{(u(t))^2 ((u'(t))^2 + (v'(t))^2)}$$

$$= \frac{u(t)u'(t)}{(u'(t))^2 + (v'(t))^2},$$

$$\Gamma_{22}^2(t,\theta) = \frac{-F(t,\theta)F_\theta(t,\theta) + \frac{1}{2}F(t,\theta)G_t(t,\theta) + \frac{1}{2}E(t,\theta)G_\theta(t,\theta)}{(W(t,\theta))^2}$$

$$= 0, \quad t \in I, \quad \theta \in [0, 2\pi].$$

Exercise 4.1. Prove that

$$\Gamma_{11}^1 + \Gamma_{12}^2 = (\log(\sqrt{EG - F^2}))_{t_1},$$

$$\Gamma_{12}^1 + \Gamma_{22}^2 = (\log(\sqrt{EG - F^2}))_{t_2}.$$

Solution. For the first equality, we have

$$(\log(\sqrt{EG - F^2}))_{t_1} = \frac{(\sqrt{EG - F^2})_{t_1}}{\sqrt{EG - F^2}}$$

$$= \frac{E_{t_1}G + EG_{t_1} - 2FF_{t_1}}{2(EG - F^2)}.$$

On the other hand,

$$\Gamma_{11}^1 + \Gamma_{12}^2 = \frac{\frac{1}{2}E_{t_1}G - F_{t_1}F + \frac{1}{2}E_{t_2}F}{W^2} + \frac{-E_{t_2}F + EG_{t_1}}{2W^2}$$

$$= \frac{E_{t_1}G + EG_{t_1} - 2FF_{t_1}}{2(EG - F^2)},$$

proving the first equality. The second equality can be proved in a similar way.

Exercise 4.2. Let $F = 0$. Find the Christoffel coefficients.

Answer 4.1.

$$\Gamma_{11}^1 = \frac{E_{t_1}}{2E},$$

$$\Gamma_{11}^2 = -\frac{E_{t_2}}{2G},$$

$$\Gamma_{12}^1 = \frac{E_{t_2}}{2E},$$

$$\Gamma_{12}^2 = \frac{G_{t_1}}{2G},$$

$$\Gamma_{22}^1 = -\frac{G_{t_1}}{2E},$$

$$\Gamma_{22}^2 = \frac{G_{t_2}}{2G}.$$

Exercise 4.3. Find a_{ij}, $i, j \in \{1, 2\}$, such that

$$n_{t_1} = a_{11}f_{t_1} + a_{12}f_{t_2},$$

$$n_{t_2} = a_{21}f_{t_1} + a_{22}f_{t_2}. \tag{4.2}$$

Definition 4.2. The coefficients a_{ij}, $i, j \in \{1, 2\}$, are called the Weingarten coefficients.

Solution. By the first equation of the system (4.2), we find

$$n_{t_1} \cdot f_{t_1} = a_{11}(f_{t_1} \cdot f_{t_1}) + a_{12}(f_{t_1} \cdot f_{t_2}),$$
$$n_{t_1} \cdot f_{t_2} = a_{11}(f_{t_1} \cdot f_{t_2}) + a_{12}(f_{t_2} \cdot f_{t_2}),$$

or

$$a_{11}E + a_{12}F = -L,$$
$$a_{11}F + a_{12}G = -M.$$

Therefore

$$a_{11} = \frac{FM - GL}{W^2},$$
$$a_{12} = \frac{FL - EM}{W^2}.$$

Using the second equation of the system (4.2), we arrive at

$$n_{t_2} \cdot f_{t_1} = a_{21}(f_{t_1} \cdot f_{t_1}) + a_{22}(f_{t_1} \cdot f_{t_2}),$$
$$n_{t_2} \cdot f_{t_2} = a_{21}(f_{t_1} \cdot f_{t_2}) + a_{22}(f_{t_2} \cdot f_{t_2}),$$

or

$$a_{21}E + a_{22}F = -M,$$
$$a_{21}F + a_{22}G = -N.$$

Consequently,

$$a_{21} = \frac{-GM + FN}{W^2},$$
$$a_{22} = \frac{FM - EN}{W^2}.$$

Exercise 4.4. Prove that

$$(\Gamma_{11}^1)_{t_2} - (\Gamma_{12}^1)_{t_1} + \Gamma_{11}^2 \Gamma_{22}^1 - \Gamma_{12}^2 \Gamma_{12}^1 = -a_{21}L + a_{11}M,$$
$$(\Gamma_{11}^2)_{t_2} - (\Gamma_{12}^2)_{t_1} + \Gamma_{11}^1 \Gamma_{12}^2 + \Gamma_{11}^2 \Gamma_{22}^2 - \Gamma_{12}^1 \Gamma_{11}^2 - (\Gamma_{12}^2)^2 = Ma_{12} - La_{22},$$
$$(\Gamma_{12}^1)_{t_2} - (\Gamma_{22}^1)_{t_1} + (\Gamma_{12}^1)^2 + \Gamma_{12}^2 \Gamma_{22}^1 - \Gamma_{22}^1 \Gamma_{11}^1 - \Gamma_{22}^2 \Gamma_{12}^1 = a_{11}N - Ma_{21},$$
$$(\Gamma_{12}^2)_{t_2} - (\Gamma_{22}^2)_{t_1} + \Gamma_{12}^1 \Gamma_{12}^2 + \Gamma_{12}^2 \Gamma_{22}^2 - \Gamma_{22}^1 \Gamma_{11}^2 - \Gamma_{22}^2 \Gamma_{12}^2 = a_{12}N - a_{22}M,$$

$$(4.3)$$

and

$$L_{t_2} - M_{t_1} = -\Gamma_{11}^1 M - \Gamma_{11}^2 N + \Gamma_{12}^1 L + \Gamma_{12}^2 M,$$
$$M_{t_2} - N_{t_1} = -\Gamma_{12}^1 M - \Gamma_{12}^2 N + \Gamma_{22}^1 L + \Gamma_{22}^2 M. \tag{4.4}$$

Definition 4.3. The equations (4.3) are called the Gauss equations.

Definition 4.4. The equations (4.4) are called Codazzi–Mainardi equations.

Solution. To deduce the Gauss and Codazzi–Mainardi equations, we will use the relations

$$f_{t_1^2 t_2} = f_{t_1 t_2 t_1} \tag{4.5}$$

and

$$f_{t_1 t_2^2} = f_{t_2^2 t_1}. \tag{4.6}$$

We have

$$f_{t_1^2} = \Gamma_{11}^1 f_{t_1} + \Gamma_{11}^2 f_{t_2} + Ln$$

and then

$$\begin{aligned}
f_{t_1^2 t_2} &= (\Gamma_{11}^1)_{t_2} f_{t_1} + \Gamma_{11}^1 f_{t_1 t_2} + (\Gamma_{11}^2)_{t_2} f_{t_2} + \Gamma_{11}^2 f_{t_2 t_2} + L_{t_2} n + Ln_{t_2} \\
&= (\Gamma_{11}^1)_{t_2} f_{t_1} + (\Gamma_{11}^2)_{t_2} f_{t_2} + L_{t_2} n + \Gamma_{11}^1 (\Gamma_{12}^1 f_{t_1} + \Gamma_{12}^2 f_{t_2} + Mn) \\
&\quad + \Gamma_{11}^2 (\Gamma_{22}^1 f_{t_1} + \Gamma_{22}^2 f_{t_2} + Nn) + L(a_{21} f_{t_1} + a_{22} f_{t_2}) \\
&= ((\Gamma_{11}^1)_{t_2} + \Gamma_{11}^1 \Gamma_{12}^1 + \Gamma_{11}^2 \Gamma_{22}^1 + La_{21}) f_{t_1} \\
&\quad + ((\Gamma_{11}^2)_{t_2} + \Gamma_{11}^1 \Gamma_{12}^2 + \Gamma_{11}^2 \Gamma_{22}^2 + La_{22}) f_{t_2} + (L_{t_2} + \Gamma_{11}^1 M + \Gamma_{11}^2 N)n,
\end{aligned}$$

i. e.,

$$\begin{aligned}
f_{t_1^2 t_2} &= ((\Gamma_{11}^1)_{t_2} + \Gamma_{11}^1 \Gamma_{12}^1 + \Gamma_{11}^2 \Gamma_{22}^1 + La_{21}) f_{t_1} \\
&\quad + ((\Gamma_{11}^2)_{t_2} + \Gamma_{11}^1 \Gamma_{12}^2 + \Gamma_{11}^2 \Gamma_{22}^2 + La_{22}) f_{t_2} + (L_{t_2} + \Gamma_{11}^1 M + \Gamma_{11}^2 N)n. \tag{4.7}
\end{aligned}$$

Now, using that

$$f_{t_1 t_2} = \Gamma_{12}^1 f_{t_1} + \Gamma_{12}^2 f_{t_2} + Mn,$$

we obtain

$$\begin{aligned}
f_{t_1 t_2 t_1} &= (\Gamma_{12}^1)_{t_1} f_{t_1} + \Gamma_{12}^1 f_{t_1^2} + (\Gamma_{12}^2)_{t_1} f_{t_2} + \Gamma_{12}^2 f_{t_1 t_2} + M_{t_1} n + Mn_{t_1} \\
&= (\Gamma_{12}^1)_{t_1} f_{t_1} + (\Gamma_{12}^2)_{t_1} f_{t_2} + M_{t_1} n + \Gamma_{12}^1 (\Gamma_{11}^1 f_{t_1} + \Gamma_{11}^2 f_{t_2} + Ln) \\
&\quad + \Gamma_{12}^2 (\Gamma_{12}^1 f_{t_1} + \Gamma_{12}^2 f_{t_2} + Mn) + M(a_{11} f_{t_1} + a_{12} f_{t_2})
\end{aligned}$$

$$= ((\Gamma_{12}^1)_{t_1} + \Gamma_{12}^1\Gamma_{11}^1 + \Gamma_{12}^2\Gamma_{12}^1 + a_{11}M)f_{t_1}$$
$$+ ((\Gamma_{12}^2)_{t_1} + \Gamma_{12}^1\Gamma_{11}^2 + (\Gamma_{12}^2)^2 + a_{12}M)f_{t_2} + (M_{t_1} + \Gamma_{12}^1 L + \Gamma_{12}^2 M)n,$$

i. e.,

$$f_{t_1 t_2 t_1} = ((\Gamma_{12}^1)_{t_1} + \Gamma_{12}^1\Gamma_{11}^1 + \Gamma_{12}^2\Gamma_{12}^1 + a_{11}M)f_{t_1}$$
$$+ ((\Gamma_{12}^2)_{t_1} + \Gamma_{12}^1\Gamma_{11}^2 + (\Gamma_{12}^2)^2 + a_{12}M)f_{t_2} + (M_{t_1} + \Gamma_{12}^1 L + \Gamma_{12}^2 M)n.$$

Using the latter equation with (4.5) and (4.7), we find the first two of Gauss equations and the first of the Codazzi–Mainardi equations. Next,

$$f_{t_1 t_2^2} = (\Gamma_{12}^1)_{t_2} f_{t_1} + \Gamma_{12}^1 f_{t_1 t_2} + (\Gamma_{12}^2)_{t_2} f_{t_2} + \Gamma_{12}^2 f_{t_2 t_2} + M_{t_2} n + M n_{t_2}$$
$$= (\Gamma_{12}^1)_{t_2} f_{t_1} + (\Gamma_{12}^2)_{t_2} f_{t_2} + M_{t_2} n + \Gamma_{12}^1 (\Gamma_{12}^1 f_{t_1} + \Gamma_{12}^2 f_{t_2} + Mn)$$
$$+ \Gamma_{12}^2 (\Gamma_{22}^1 f_{t_1} + \Gamma_{22}^2 f_{t_2} + Nn) + M(a_{21}f_{t_1} + a_{22}f_{t_2})$$
$$= ((\Gamma_{12}^1)_{t_2} + \Gamma_{12}^1\Gamma_{12}^1 + (\Gamma_{12}^2)^2 + \Gamma_{12}^2\Gamma_{22}^1 + Ma_{21})f_{t_1}$$
$$+ ((\Gamma_{12}^2)_{t_2} + \Gamma_{12}^1\Gamma_{12}^2 + \Gamma_{12}^2\Gamma_{22}^2 + Ma_{22})f_{t_2} + (M_{t_2} + \Gamma_{12}^1 M + \Gamma_{12}^2 N)n,$$

i. e.,

$$f_{t_1 t_2^2} = ((\Gamma_{12}^1)_{t_2} + \Gamma_{12}^1\Gamma_{12}^1 + (\Gamma_{12}^2)^2 + \Gamma_{12}^2\Gamma_{22}^1 + Ma_{21})f_{t_1}$$
$$+ ((\Gamma_{12}^2)_{t_2} + \Gamma_{12}^1\Gamma_{12}^2 + \Gamma_{12}^2\Gamma_{22}^2 + Ma_{22})f_{t_2} + (M_{t_2} + \Gamma_{12}^1 M + \Gamma_{12}^2 N)n. \tag{4.8}$$

Now, using that

$$f_{t_2 t_2} = \Gamma_{22}^1 f_{t_1} + \Gamma_{22}^2 f_{t_2} + Nn,$$

we arrive at

$$f_{t_2^2 t_1} = (\Gamma_{22}^1)_{t_1} f_{t_1} + \Gamma_{22}^1 f_{t_1^2} + (\Gamma_{22}^2)_{t_1} f_{t_2} + \Gamma_{22}^2 f_{t_1 t_2} + N_{t_1} n + N n_{t_1}$$
$$= (\Gamma_{22}^1)_{t_1} f_{t_1} + (\Gamma_{22}^2)_{t_1} f_{t_2} + N_{t_1} n + \Gamma_{22}^1 (\Gamma_{11}^1 f_{t_1} + \Gamma_{11}^2 f_{t_2} + Ln)$$
$$+ \Gamma_{22}^2 (\Gamma_{12}^1 f_{t_1} + \Gamma_{12}^2 f_{t_2} + Mn) + N(a_{11}f_{t_1} + a_{12}f_{t_2})$$
$$= ((\Gamma_{22}^1)_{t_1} + \Gamma_{22}^1\Gamma_{11}^1 + \Gamma_{22}^2\Gamma_{12}^1 + Na_{11})f_{t_1}$$
$$+ ((\Gamma_{22}^2)_{t_1} + \Gamma_{22}^1\Gamma_{11}^2 + \Gamma_{22}^2\Gamma_{12}^2 + a_{12}N)f_{t_2} + (N_{t_2} + \Gamma_{22}^1 L + \Gamma_{22}^2 M)n,$$

i. e.,

$$f_{t_2^2 t_1} = ((\Gamma_{22}^1)_{t_1} + \Gamma_{22}^1\Gamma_{11}^1 + \Gamma_{22}^2\Gamma_{12}^1 + Na_{11})f_{t_1}$$
$$+ ((\Gamma_{22}^2)_{t_1} + \Gamma_{22}^1\Gamma_{11}^2 + \Gamma_{22}^2\Gamma_{12}^2 + a_{12}N)f_{t_2} + (N_{t_1} + \Gamma_{22}^1 L + \Gamma_{22}^2 M)n.$$

Now, applying (4.6), (4.8), and the latter expression, we find the third and fourth of Gauss equations and the second of Codazzi–Mainardi equations. This completes the proof.

4.3 The fundamental theorem in the theory of surfaces

Suppose that $U \subset \mathbb{R}^n$ is an open set. There are two symmetric matrices on U given by

$$\begin{pmatrix} g_{11} & g_{12} \\ g_{12} & g_{22} \end{pmatrix} \quad \text{and} \quad \begin{pmatrix} h_{11} & h_{12} \\ h_{12} & h_{22} \end{pmatrix}$$

of classes C^2 and C^1, respectively. Here, for any $(t_1, t_2) \in U$, the quadratic for an associated bilinear form whose matrix is (g_{ij}), $i, j \in \{1, 2\}$, is positive definite and, moreover, its components verify the Gauss and Codazzi–Mainardi compatibility conditions. Choose $q_0 = (t_1^0, t_2^0) \in U, f(q_0) = p_0 \in \mathbb{R}^3$ and the vectors

$$f_{t_1}(q_0), f_{t_2}(q_0), n(q_0) \in T_{p_0}\mathbb{R}^3$$

such that

$$f_{t_1}(q_0) \cdot f_{t_1}(q_0) = g_{11}(q_0),$$
$$f_{t_1}(q_0) \cdot f_{t_2}(q_0) = g_{12}(q_0),$$
$$f_{t_2}(q_0) \cdot f_{t_2}(q_0) = g_{22}(q_0),$$
$$n(q_0) \cdot f_{t_1}(q_0) = 0,$$
$$n(q_0) \cdot f_{t_2}(q_0) = 0,$$
$$n(q_0) \cdot n(q_0) = 1$$

and $\{f_{t_1}(q_0), f_{t_2}(q_0), n(q_0)\}$ is a right-handed basis of the vector space $T_{p_0}\mathbb{R}^3$. Set

$$\Gamma_{11}^1 = \frac{(g_{11})_{t_1}g_{22} - 2(g_{12})_{t_1}g_{12} + (g_{11})_{t_2}g_{12}}{2(g_{11}g_{22} - g_{12}^2)},$$

$$\Gamma_{11}^2 = \frac{-(g_{11})_{t_1}g_{12} + 2g_{11}(g_{12})_{t_1} - g_{11}(g_{11})_{t_2}}{2(g_{11}g_{22} - g_{12}^2)},$$

$$\Gamma_{12}^1 = \frac{(g_{11})_{t_2}g_{22} - (g_{22})_{t_1}g_{12}}{2(g_{11}g_{22} - g_{12}^2)},$$

$$\Gamma_{12}^2 = \frac{-(g_{11})_{t_2}g_{12} + g_{11}(g_{22})_{t_1}}{2(g_{11}g_{22} - g_{12}^2)},$$

$$\Gamma_{21}^1 = \Gamma_{12}^1,$$

$$\Gamma_{21}^2 = \Gamma_{12}^2,$$

$$\Gamma_{22}^1 = \frac{2(g_{12})_{t_2}g_{22} - (g_{22})_{t_1}g_{22} - (g_{22})_{t_2}g_{12}}{2(g_{11}g_{22} - g_{12}^2)},$$

$$\Gamma_{22}^2 = \frac{-2g_{12}(g_{12})_{t_2} + g_{12}(g_{22})_{t_1} + g_{11}(g_2)_{t_2}}{2(g_{11}g_{22} - g_{12}^2)}.$$

Consider the Cauchy problem

$$\frac{\partial f_{t_1}}{\partial t_1} = \Gamma_{11}^1 r_{t_1} + \Gamma_{11}^2 f_{t_2} + h_{11} n,$$

$$\frac{\partial f_{t_1}}{\partial t_2} = \Gamma_{12}^1 r_{t_1} + \Gamma_{12}^2 f_{t_2} + h_{12} n,$$

$$\frac{\partial f_{t_2}}{\partial t_1} = \Gamma_{21}^1 r_{t_1} + \Gamma_{21}^2 f_{t_2} + h_{21} n,$$

$$\frac{\partial n}{\partial t_1} = \frac{1}{W^2}((g_{12}h_{12} - g_{22}h_{11})f_{t_1} + (g_{12}h_{11} - g_{11}h_{12})f_{t_2}), \qquad (4.9)$$

$$\frac{\partial n}{\partial t_2} = \frac{1}{W^2}((g_{21}h_{22} - g_{22}h_{21})f_{t_1} - (g_{12}h_{21} - g_{11}h_{22})f_{t_2}),$$

$$r_{t_1}(q_0) = f_{t_1}^0,$$

$$r_{t_2}(q_0) = f_{t_2}^0,$$

$$n(q_0) = n_0.$$

Here

$$W^2 = g_{11}g_{22} - g_{12}^2.$$

Exercise 4.5. Prove that:

1.
$$\frac{\partial^2 f_{t_1}}{\partial t_1 \partial t_2} = \frac{\partial^2 f_{t_1}}{\partial t_2 \partial t_1};$$

2.
$$\frac{\partial^2 f_{t_2}}{\partial t_1 \partial t_2} = \frac{\partial^2 f_{t_2}}{\partial t_2 \partial t_1};$$

3.
$$\frac{\partial^2 n}{\partial t_1 \partial t_2} = \frac{\partial^2 n}{\partial t_2 \partial t_1}.$$

Solution. 1. We have

$$\frac{\partial^2 f_{t_1}}{\partial t_2 \partial t_1} = \frac{\partial f_{t_1 t_1}}{\partial t_2} = \frac{\partial}{\partial t_2}(\Gamma_{11}^1 f_{t_1} + \Gamma_{11}^2 f_{t_2} + h_{11} n)$$

$$= \frac{\partial \Gamma_{11}^1}{\partial t_2} f_{t_1} + \Gamma_{11}^1 \frac{\partial f_{t_1}}{\partial t_2} + \frac{\partial \Gamma_{11}^2}{\partial t_2} f_{t_2} + \Gamma_{11}^2 \frac{\partial f_{t_2}}{\partial t_2} + \frac{\partial h_{11}}{\partial t_2} n + h_{11} \frac{\partial n}{\partial t_2}$$

$$= \frac{\partial \Gamma_{11}^1}{\partial t_2} f_{t_1} + \Gamma_{11}^1 (\Gamma_{12}^1 f_{t_1} + \Gamma_{12}^2 f_{t_2} + h_{12} n)$$

$$+ \frac{\partial \Gamma_{11}^2}{\partial t_2} f_{t_2} + \Gamma_{11}^2 (\Gamma_{22}^1 f_{t_1} + \Gamma_{22}^2 f_{t_2} + h_{22} n)$$

$$+ \frac{\partial h_{11}}{\partial t_2} n + \frac{1}{W^2} h_{11} ((g_{21} h_{22} - g_{22} h_{11}) f_{t_1} + (g_{12} h_{11} - g_{11} h_{12}) f_{t_2})$$

$$= \left(\frac{\partial \Gamma_{11}^1}{\partial t_2} + \Gamma_{11}^1 \Gamma_{12}^1 + \Gamma_{11}^2 \Gamma_{22}^1 + \frac{h_{11}}{W^2} (g_{21} h_{22} - g_{22} h_{12}) \right) f_{t_1}$$

$$+ \left(\frac{\partial \Gamma_{11}^2}{\partial t_2} + \Gamma_{11}^1 \Gamma_{12}^2 + \Gamma_{11}^2 \Gamma_{22}^2 + \frac{h_{11}}{W^2} (g_{12} h_{12} - g_{11} h_{22}) \right) f_{t_2}$$

$$+ \left(\Gamma_{11}^1 h_{12} + \Gamma_{11}^2 h_{22} + \frac{\partial h_{11}}{\partial t_2} \right) n$$

$$= A_1 f_{t_1} + A_2 f_{t_2} + A_3 n$$

and

$$\frac{\partial^2 f_{t_1}}{\partial t_1 \partial t_2} = \frac{\partial f_{t_1 t_2}}{\partial t_1} = \frac{\partial}{\partial t_1} (\Gamma_{12}^1 f_{t_1} + \Gamma_{12}^2 f_{t_2} + h_{12} n)$$

$$= \frac{\Gamma_{12}^1}{\partial t_1} f_{t_1} + \Gamma_{12}^1 \frac{\partial f_{t_1}}{\partial t_1} + \frac{\partial \Gamma_{12}^2}{\partial t_1} f_{t_2} + \Gamma_{12}^2 \frac{\partial f_{t_2}}{\partial t_1} + \frac{\partial h_{12}}{\partial t_1} n + h_{12} \frac{\partial n}{\partial t_1}$$

$$= \frac{\partial \Gamma_{12}^1}{\partial t_1} f_{t_1} + \Gamma_{12}^1 (\Gamma_{11}^1 f_{t_1} + \Gamma_{11}^2 f_{t_2} + h_{11} n)$$

$$+ \frac{\partial \Gamma_{12}^2}{\partial t_1} f_{t_2} + \Gamma_{12}^2 (\Gamma_{21}^1 f_{t_1} + \Gamma_{21}^2 f_{t_2} + h_{21} n) + \frac{\partial h_{12}}{\partial t_1} n$$

$$+ \frac{h_{12}}{W^2} ((g_{12} h_{12} - g_{22} h_{11}) f_{t_1} + (g_{12} h_{11} - g_{11} h_{12}) f_{t_2})$$

$$= \left(\frac{\Gamma_{12}^1}{\partial t_1} + \Gamma_{12}^1 \Gamma_{11}^1 + \Gamma_{12}^2 \Gamma_{21}^1 + \frac{h_{12}}{W^2} (g_{12} h_{12} - g_{22} h_{11}) \right) f_{t_1}$$

$$+ \left(\frac{\partial \Gamma_{12}^2}{\partial t_1} + \Gamma_{12}^1 \Gamma_{11}^2 + \Gamma_{12}^2 \Gamma_{21}^2 + \frac{h_{12}}{W^2} (g_{12} h_{11} - g_{11} h_{12}) \right) f_{t_2}$$

$$+ \left(\Gamma_{12}^1 h_{11} + \frac{\partial h_{12}}{\partial t_1} + \Gamma_{12}^2 h_{21} \right) n$$

$$= B_1 f_{t_1} + B_2 f_{t_2} + B_3 n.$$

Now, applying the Gauss equations, we find

$$A_1 - B_1 = \frac{\partial \Gamma_{11}^1}{\partial t_2} + \Gamma_{11}^1 \Gamma_{12}^1 + \Gamma_{11}^2 \Gamma_{22}^1 + \frac{h_{11}}{W^2} (g_{21} h_{22} - g_{22} h_{21})$$

$$- \frac{\partial \Gamma_{12}^1}{\partial t_1} - \Gamma_{12}^1 \Gamma_{11}^1 - \Gamma_{12}^2 \Gamma_{21}^1 - \frac{h_{12}}{W^2} (g_{12} h_{12} - g_{22} h_{11})$$

$$= \frac{\partial \Gamma_{11}^1}{\partial t_2} - \frac{\partial \Gamma_{12}^1}{\partial t_1} + \Gamma_{11}^2 \Gamma_{22}^1 - \Gamma_{12}^2 \Gamma_{21}^1 + g_{21} \frac{1}{W^2} (h_{11} h_{22} - h_{12}^2)$$

$$= -\frac{h_{11}}{W^2}(-g_{22}h_{12} + g_{12}h_{22}) + \frac{h_{12}}{W^2}(g_{12}h_{12} - g_{22}h_{11})$$

$$+ g_{21}\frac{1}{W^2}(h_{11}h_{22} - h_{12}^2)$$

$$= -g_{12}\frac{1}{W^2}(h_{11}h_{22} - h_{12}^2) + g_{21}\frac{1}{W^2}(h_{11}h_{22} - h_{12}^2)$$

$$= 0$$

and

$$A_2 - B_2 = \frac{\partial \Gamma_{11}^2}{\partial t_2} + \Gamma_{11}^1\Gamma_{12}^2 + \Gamma_{11}^2\Gamma_{22}^2 + \frac{h_{11}}{W^2}(g_{12}h_{12} - g_{11}h_{22})$$

$$- \frac{\partial \Gamma_{12}^2}{\partial t_1} - \Gamma_{12}^1\Gamma_{11}^2 - \Gamma_{12}^2\Gamma_{21}^2 - \frac{h_{12}}{W^2}(g_{12}h_{11} - g_{11}h_{12})$$

$$= \frac{\partial \Gamma_{11}^2}{\partial t_2} - \frac{\partial \Gamma_{12}^2}{\partial t_1} + \Gamma_{11}^1\Gamma_{12}^2 + \Gamma_{11}^2\Gamma_{22}^2 - \Gamma_{12}^1\Gamma_{11}^2 - \Gamma_{12}^2\Gamma_{21}^2$$

$$+ g_{11}\frac{1}{W^2}(h_{11}h_{22} - h_{12}^2)$$

$$= h_{12}\frac{1}{W^2}(g_{12}h_{11} - g_{11}h_{12}) - h_{11}\frac{1}{W^2}(g_{12}h_{12} - g_{11}h_{22})$$

$$+ g_{11}\frac{1}{W^2}(h_{11}h_{22} - h_{12}^2)$$

$$= -g_{11}\frac{1}{W^2}(h_{11}h_{22} - h_{12}^2) + g_{11}\frac{1}{W^2}(h_{11}h_{22} - h_{12}^2)$$

$$= 0.$$

Now, we apply the Codazzi–Mainardi equations and find

$$A_3 - B_3 = \Gamma_{11}^1 h_{12} + \Gamma_{11}^2 h_{22} + \frac{\partial h_{11}}{\partial t_2} - \Gamma_{12}^1 h_{11} - \frac{\partial h_{12}}{\partial t_1} - \Gamma_{12}^2 h_{21}$$

$$= \frac{\partial h_{11}}{\partial t_2} - \frac{\partial h_{12}}{\partial t_1} + \Gamma_{11}^1 h_{12} + \Gamma_{11}^2 h_{22} - \Gamma_{12}^1 h_{11} - \Gamma_{12}^2 h_{21}$$

$$= 0.$$

Therefore

$$\frac{\partial^2 f_{t_1}}{\partial t_1 \partial t_2} = \frac{\partial^2 f_{t_1}}{\partial t_2 \partial t_1}.$$

2.

Hint 4.2. Use the solution of part 1.

3.

Hint 4.3. Use the solution of part 1.

Exercise 4.6. Prove that there exists an open neighborhood $Q \subset U$ of the point q_0 and a set of C^2 vector functions $f_{t_1}, f_{t_2}, n : W \to \mathbb{R}^3$ which are solutions to the initial value problem (IVP) (4.9).

Solution. By Exercise 4.5, it follows that the IVP is completely integrable. Hence, the desired result follows.

Exercise 4.7. Prove that the vectors

$$f_{t_1}(q_0) = (\sqrt{g_{11}(q_0)}, 0, 0),$$

$$f_{t_2}(q_0) = \left(\frac{g_{12}((q_0))}{\sqrt{g_{11}((q_0))}}, \frac{\sqrt{g_{11}(q_0)g_{22}(q_0) - (g_{12}(q_0))^2}}{\sqrt{g_{11}(q_0)}}, 0 \right),$$

$$n(q_0) = (0, 0, 1)$$

satisfy all the hypotheses.

Hint 4.4. Use that the matrix

$$\begin{pmatrix} g_{11}(q_0) & g_{12}(q_0) \\ g_{12}(q_0) & g_{22}(q_0) \end{pmatrix}$$

is positive definite.

Exercise 4.8. Consider the IVP

$$\frac{\partial f}{\partial t_1} = r_{t_1},$$

$$\frac{\partial f}{\partial t_2} = r_{t_2}, \tag{4.10}$$

$$f(q_0) = p_0.$$

Prove that there exist an open neighborhood $V \subset W \subset U$ of u^0 and a single C^3 function $f : V \to \mathbb{R}^3$ that is a solution to the IVP (4.10).

Solution. By the second and third equations of (3.2), we find

$$\frac{\partial^2 f}{\partial t_2 \partial t_1} = \frac{\partial f_{t_1}}{\partial t_2}$$

$$= \Gamma_{12}^1 f_{t_1} + \Gamma_{12}^2 f_{t_2} + h_{12} n$$

$$= \Gamma_{21}^1 f_{t_1} + \Gamma_{21}^2 f_{t_2} + h_{21} n$$

$$= \frac{\partial^2 f}{\partial t_1 \partial t_2}.$$

Thus, (4.10) is completely integrable and the desired result follows.

Exercise 4.9. Prove that the surface f in Exercise 4.8 satisfies the following relations:

1.
$$\frac{\partial(f_{t_1} \cdot f_{t_1})}{\partial t_1} = 2(\Gamma_{11}^1(r_{t_1} \cdot r_{t_1}) + \Gamma_{11}^2(r_{t_1} \cdot r_{t_2}) + h_{11}(r_{t_1} \cdot n));$$

2.
$$\frac{\partial(f_{t_1} \cdot f_{t_1})}{\partial t_2} = 2(\Gamma_{12}^1(f_{t_1} \cdot f_{t_1}) + \Gamma_{12}^2(f_{t_1} \cdot f_{t_2}) + h_{12}(f_{t_1} \cdot n));$$

3.
$$\frac{\partial(f_{t_1} \cdot f_{t_2})}{\partial t_1} = \Gamma_{12}^1(f_{t_1} \cdot f_{t_1}) + (\Gamma_{11}^1 + \Gamma_{12}^2)(f_{t_1} \cdot f_{t_2}) + \Gamma_{11}^2(f_{t_2} \cdot f_{t_2})$$
$$+ h_{12}(n \cdot f_{t_1}) + h_{11}(n \cdot f_{t_2});$$

4.
$$\frac{\partial(f_{t_1} \cdot f_{t_2})}{\partial t_2} = \Gamma_{22}^1(f_{t_1} \cdot f_{t_1}) + (\Gamma_{12}^1 + \Gamma_{22}^2)(f_{t_1} \cdot f_{t_2}) + \Gamma_{12}^2(f_{t_2} \cdot f_{t_2})$$
$$+ h_{22}(n \cdot f_{t_1}) + h_{12}(n \cdot f_{t_2});$$

5.
$$\frac{\partial(f_{t_2} \cdot f_{t_2})}{\partial t_1} = 2(\Gamma_{12}^1(f_{t_1} \cdot f_{t_2}) + \Gamma_{12}^2(f_{t_2} \cdot f_{t_2}) + h_{12}(n \cdot f_{t_2}));$$

6.
$$\frac{\partial(f_{t_2} \cdot f_{t_2})}{\partial t_2} = 2(\Gamma_{22}^1(f_{t_1} \cdot f_{t_2}) + \Gamma_{22}^2(f_{t_2} \cdot f_{t_2}) + h_{22}(n \cdot f_{t_2}));$$

7.
$$\frac{\partial(f_{t_1} \cdot n)}{\partial t_1} = \Gamma_{11}^1(f_{t_1} \cdot n) + \Gamma_{11}^2(f_{t_2} \cdot n) + h_{11}(n \cdot n)$$
$$- \frac{1}{W^2}(h_{11}g_{11} + h_{12}g_{12})(f_{t_1} \cdot f_{t_1}) - \frac{1}{W^2}(h_{11}g_{12} + h_{12}g_{22})(f_{t_1} \cdot f_{t_2});$$

8.
$$\frac{\partial(f_{t_1} \cdot n)}{\partial t_2} = \Gamma_{12}^1(f_{t_1} \cdot n) + \Gamma_{12}^2(f_{t_2} \cdot n) + h_{12}(n \cdot n)$$
$$- \frac{1}{W^2}(h_{21}g_{11} + h_{22}g_{12})(f_{t_1} \cdot f_{t_1}) - \frac{1}{W^2}(h_{21}g_{12} + h_{22}g_{22})(f_{t_1} \cdot f_{t_2});$$

9.
$$\frac{\partial(f_{t_2} \cdot n)}{\partial t_1} = \Gamma_{12}^1(f_{t_1} \cdot n) + \Gamma_{12}^2(f_{t_2} \cdot n) + h_{12}(n \cdot n)$$
$$- \frac{1}{W^2}(h_{11}g_{11} + h_{12}g_{12})(f_{t_1} \cdot f_{t_2}) - \frac{1}{W^2}(h_{11}g_{12} + h_{12}g_{22})(f_{t_1} \cdot f_{t_2});$$

10.
$$\frac{\partial(f_{t_2} \cdot n)}{\partial t_2} = \Gamma_{22}^1(f_{t_1} \cdot n) + \Gamma_{22}^2(f_{t_2} \cdot n) + h_{22}(n \cdot n)$$
$$- \frac{1}{W^2}(h_{21}g_{11} + h_{22}g_{12})(f_{t_1} \cdot f_{t_1}) - \frac{1}{W^2}(h_{21}g_{12} + h_{22}g_{22})(f_{t_1} \cdot f_{t_2});$$

11.
$$\frac{\partial(n \cdot n)}{\partial t_1} = \frac{2}{W^2}(-(h_{11}g_{11} + h_{12}g_{12})(f_{t_1} \cdot n) - (h_{11}g_{12} + h_{12}g_{22})(f_{t_2} \cdot n));$$

12.
$$\frac{\partial(n \cdot n)}{\partial t_2} = \frac{2}{W^2}(-(h_{21}g_{11} + h_{22}g_{12})(f_{t_1} \cdot n) - (h_{21}g_{12} + h_{22}g_{22})(f_{t_2} \cdot n)).$$

1.

Solution. Using the Gauss–Weingarten equations, we have

$$f_{t_1 t_1} = \Gamma_{11}^1 f_{t_1} + \Gamma_{11}^2 f_{t_2} + h_{11}n.$$

Then

$$\frac{\partial(f_{t_1} \cdot f_{t_1})}{\partial t_1} = (f_{t_1 t_1} \cdot f_{t_1}) + (f_{t_1} \cdot f_{t_1 t_1})$$

$$= 2(f_{t_1 t_1} \cdot f_{t_1})$$

$$= 2(\Gamma_{11}^1 f_{t_1} + \Gamma_{11}^2 f_{t_2} + h_{11}n) \cdot f_{t_1}$$

$$= 2(\Gamma_{11}^1 (f_{t_1} \cdot f_{t_1}) + \Gamma_{11}^2 (f_{t_1} \cdot f_{t_2}) + h_{11}(n \cdot r_{t_1})).$$

Hint 4.5. For parts 2–12, use the solution of part 1.

Exercise 4.10. Prove that

$$f_{t_1} \cdot f_{t_1} = g_{11},$$
$$f_{t_1} \cdot f_{t_2} = g_{12},$$
$$f_{t_1} \cdot f_{t_2} = g_{22},$$
$$f_{t_1} \cdot n = 0,$$
$$f_{t_2} \cdot n = 0,$$
$$n \cdot n = 1$$

satisfy all the relations in Exercise 4.9.

Hint 4.6. Use the definition of Γ_{ij}^k, $i, j, k \in \{1, 2\}$.

Exercise 4.11. Let f be the surface in Exercise 4.8. Prove that the matrix

$$\begin{pmatrix} g_{11} & g_{12} \\ g_{12} & g_{22} \end{pmatrix}$$

is the matrix of its first fundamental form.

Hint 4.7. Use Exercise 4.10.

Exercise 4.12. Let f be the surface in Exercise 4.8. Prove that the matrix

$$\begin{pmatrix} h_{11} & h_{12} \\ h_{12} & h_{22} \end{pmatrix}$$

is the matrix of its second fundamental form.

Hint 4.8. Use Exercise 4.11.

Exercise 4.13 (Bonnet theorem). Prove that there exists a single regular surface of class C^3, $f : V \to \mathbb{R}^3$, with $V \subset U$ an open set such that the following conditions hold:

1. $f(q_0) = p_0$.
2. $f_{t_1}(q_0) = f_{t_1}^0$, $f_{t_2}(q_0) = f_{t_2}^0$.
3. $n(q_0) = n^0$.
4. g_{ij} and h_{ij}, $i, j \in \{1, 2\}$, are the elements of the matrices of the first and second fundamental forms of f, respectively.

Hint 4.9. Use Exercises 4.5–4.12.

4.4 The Darboux frame

Let S be an oriented surface and (U, f) be a local representation of S that is compatible with the orientation of S, and (I, g) be a parameterized curve whose support lies on S. Let also $\mathbf{t}$ be the unit tangent vector of g, and n be the unit normal of S. Take

$$N = n \times \mathbf{t}.$$

Definition 4.5. The frame $\{\mathbf{t}, n, N\}$ is called the Darboux frame of S along the curve g.

Exercise 4.14. Express the vectors n and N in the terms of the Frenet frame $\{\mathbf{t}, \mathbf{n}, \mathbf{b}\}$ of g.

Solution. Let

$$\theta = \angle(\mathbf{n}, n).$$

We have that

$$\mathbf{n} = \cos(N, \mathbf{n})N + \sin(N, \mathbf{n})n,$$
$$\mathbf{b} = \cos(N, \mathbf{b})N + \sin(N, \mathbf{b})n.$$

Since

$$\angle(N, \mathbf{n}) = \frac{\pi}{2} - \theta,$$
$$\angle(N, \mathbf{b}) = \pi - \theta,$$

we find

$$\mathbf{n} = \sin \theta N + \cos \theta n,$$
$$\mathbf{b} = -\cos \theta N + \sin \theta n,$$

and

$$N = \sin \theta \mathbf{n} - \cos \theta \mathbf{b},$$
$$n = \cos \theta \mathbf{n} + \sin \theta \mathbf{b}.$$

Exercise 4.15. Prove that:
1. $\mathbf{t}' \cdot n = -\mathbf{t} \cdot n'$;
2. $\mathbf{t}' \cdot N = -\mathbf{t} \cdot N'$;
3. $n \cdot N' = -n' \cdot N$.

Here the differentiation is with respect to the arc length parameter s of the curve.

Solution. 1.   We have

$$\mathbf{t} \cdot n = 0.$$

Differentiating with respect to s, we find

$$0 = \mathbf{t}' \cdot n + \mathbf{t} \cdot n',$$

whereupon we get the desired result.

2.

Hint 4.10. Use the solution of part 1.

3.

Hint 4.11. Use the solution of part 1.

Exercise 4.16. Find representations of $\mathbf{t}'$, N', and n' in the terms of the Darboux frame $\{\mathbf{t}, n, N\}$ with respect to the arc length of the curve.

Solution. It is obvious that the derivative of a vector of the Darboux frame is perpendicular to itself. Then

$$\mathbf{t}' = a(s)N + b(s)n,$$
$$N' = c(s)\mathbf{t} + d(s)n, \qquad\qquad (4.11)$$
$$n' = e(s)\mathbf{t} + f(s)N,$$

where a, b, c, d, e, f are smooth functions of the arc length parameter s. Using the first and second equations of (4.11), we find

$$
\begin{aligned}
\mathbf{t}' \cdot N &= (a(s)N + b(s)n) \cdot N \\
&= a(s) \\
&= -\mathbf{t} \cdot N' \\
&= -\mathbf{t} \cdot (c(s)\mathbf{t} + d(s)n) \\
&= -c(s),
\end{aligned}
$$

i. e.,

$$a(s) = -c(s).$$

From the first and third equations of (4.11), we obtain

$$
\begin{aligned}
\mathbf{t}' \cdot n &= (a(s)N + b(s)n) \cdot n \\
&= b(s) \\
&= -\mathbf{t} \cdot n' \\
&= -\mathbf{t} \cdot (e(s)\tau + f(s)n) \\
&= -e(s),
\end{aligned}
$$

i. e.,

$$b(s) = -e(s).$$

From the second and third equations of (4.11), we obtain

$$
\begin{aligned}
N' \cdot n &= (c(s)\mathbf{t} + d(s)n) \cdot n \\
&= d(s) \\
&= -N \cdot n' \\
&= -N \cdot (e(s)\mathbf{t} + f(s)N) \\
&= -f(s),
\end{aligned}
$$

i. e.,

$$d(s) = -f(s).$$

Thus, the system (4.11) can rewritten in the form

$$
\begin{aligned}
\tau' &= a(s)N + b(s)n, \\
N' &= -a(s)\tau + d(s)n, \\
n' &= -b(s)\tau - d(s)N.
\end{aligned}
$$

We have, using the Frenet formulae,

$$
\begin{aligned}
\mathbf{t}' &= \kappa\mathbf{n} \\
&= \kappa \sin\theta N + \kappa \cos\theta n.
\end{aligned}
$$

Therefore

$$a(s) = \kappa \sin\theta,$$

$$b(s) = \kappa \cos \theta.$$

Now, we differentiate the equation

$$N = \sin \theta \mathbf{n} - \cos \theta \mathbf{b}$$

with respect to the arc length parameter and, using the Frenet formula, arrive at

$$
\begin{aligned}
N' &= \theta' \cos \theta \mathbf{n} + \sin \theta \mathbf{n}' + \theta' \sin \theta \mathbf{b} - \cos \theta \mathbf{b} \\
&= \theta' \cos \theta \mathbf{n} + \sin \theta(-\kappa \mathbf{t} + \tau \mathbf{b}) + \theta' \sin \theta \mathbf{b} + \tau \cos \theta \mathbf{n} \\
&= -\kappa \sin \theta \mathbf{t} + (\theta' + \tau)(\cos \theta \mathbf{n} + \sin \theta \mathbf{b}) \\
&= -\kappa \sin \theta \mathbf{t} + (\theta' + \tau)n.
\end{aligned}
$$

Therefore,

$$d(s) = \theta' + \tau.$$

So, we find

$$
\begin{aligned}
\mathbf{t}' &= \kappa \sin \theta N + \kappa \cos \theta n, \\
N' &= -\kappa \sin \theta \mathbf{t} + (\theta' + \tau)n, \\
n' &= -\kappa \cos \theta \mathbf{t} - (\theta' + \tau)N.
\end{aligned}
$$

Definition 4.6. The quantity $\kappa_g = \kappa \sin \theta$ is called the geodesic curvature or tangent curvature of the line g.

Definition 4.7. The quantity $\tau_g = \theta' + \tau$ is called the geodesic torsion.

4.5 The geodesic curvature. Geodesic lines

In this section, we will deduct some expressions for the geodesic curvature. We will use the notations in the previous section.

Exercise 4.17. Suppose that g is naturally parameterized. Prove that:
1. $\kappa_g = \mathbf{t}' \cdot N = -(\mathbf{t} \cdot N')$;
2. $\kappa_g = \mathbf{t}' \cdot (n \times \mathbf{t}) = (\mathbf{t} \times \mathbf{t}') \cdot n.$

Hint 4.12. Use the definition for κ_g and the representations of $\mathbf{t}'$ and N'.

Exercise 4.18. Let $g = g(u(t), v(t))$ and s be the arc length parameter. Prove that

$$\kappa_g = \frac{1}{(s')^3}(g' \times g'') \cdot n.$$

Solution. We have

$$\mathbf{t} = \frac{d\mathbf{g}}{ds}$$
$$= \frac{d\mathbf{g}}{dt}\frac{dt}{ds}$$
$$= \frac{\mathbf{g}'}{s'}.$$

Hence,

$$\mathbf{t}' = \frac{d\mathbf{t}}{ds}$$
$$= \frac{d\mathbf{t}}{dt}\frac{dt}{ds}$$
$$= \frac{1}{s'}\frac{\mathbf{g}''s' - \mathbf{g}'s''}{(s')^2}$$
$$= \frac{\mathbf{g}''s' - \mathbf{g}'s''}{(s')^3}.$$

Now, using Exercise 4.17, we get

$$\kappa_g = \left(\frac{\mathbf{g}'}{s'} \times \left(\frac{\mathbf{g}''s' - \mathbf{g}'s''}{(s')^3}\right)\right) \cdot \mathbf{n}$$
$$= \frac{1}{(s')^4}(\mathbf{g}' \times (\mathbf{g}''s' - \mathbf{g}'s'')) \cdot \mathbf{n}$$
$$= \frac{1}{(s')^4}(\mathbf{g}' \times (\mathbf{g}''s')) \cdot \mathbf{n}$$
$$= \frac{1}{(s')^3}(\mathbf{g}' \times \mathbf{g}'') \cdot \mathbf{n}.$$

Exercise 4.19. Let $\mathbf{g} = \mathbf{g}(u(t), v(t))$ and s be the arc length parameter. Prove that:
1.
$$\mathbf{g}'' = (\Gamma_{11}^1(u')^2 + 2\Gamma_{12}^1 u'v' + \Gamma_{22}^1(v')^2)\mathbf{g}_u$$
$$+ (\Gamma_{11}^2(u')^2 + 2\Gamma_{12}^2 u'v' + \Gamma_{22}^2(v')^2)\mathbf{g}_v + \phi_2(\mathbf{g}',\mathbf{g}') \cdot \mathbf{n} + u''\mathbf{g}_u + v''\mathbf{g}_v,$$

where ϕ_2 is the second fundamental form of the surface.
2.
$$\kappa_g = \frac{\|\mathbf{g}_u \times \mathbf{g}_v\|}{(s')^3}(\Gamma_{11}^2(u')^3 + (2\Gamma_{12}^2 - \Gamma_{11}^1)(u')^2 v' + (\Gamma_{22}^2 - 2\Gamma_{12}^1)u'(v')^2$$
$$- \Gamma_{22}^1(v')^3 + u'v'' - v'u'');$$

3.
$$\kappa_g = \frac{\|\mathbf{g}_u \times \mathbf{g}_v\|}{(s')^3}\det\begin{pmatrix} u' & u'' + (u')^2\Gamma_{11}^1 + 2u'v'\Gamma_{12}^1 + (v')^2\Gamma_{22}^1 \\ v' & v'' + (u')^2\Gamma_{11}^2 + 2u'v'\Gamma_{12}^2 + (v')^2\Gamma_{22}^2 \end{pmatrix}.$$

Definition 4.8. The curve g is called a geodesic line if $\kappa_g = 0$.

Exercise 4.20. Let g be a geodesic line. Prove that

$$0 = \Gamma_{11}^2(u')^3 + (2\Gamma_{12}^2 - \Gamma_{11}^1)(u')^2 v' + (\Gamma_{22}^2 - 2\Gamma_{12}^1)u'(v')^2 - \Gamma_{22}^1(v')^3 + u'v'' - v'u''.$$

Hint 4.13. Use Exercise 4.19.

Exercise 4.21. Let g be a geodesic line. Prove that

$$u'' + (u')^2\Gamma_{11}^1 + 2u'v'\Gamma_{12}^1 + (v')^2\Gamma_{22}^1 = 0,$$
$$v'' + (u')^2\Gamma_{11}^2 + 2u'v'\Gamma_{12}^2 + (v')^2\Gamma_{22}^2 = 0.$$

4.6 Geodesics of Liouville surfaces

Definition 4.9. A surface S is called a Liouville surface if its first fundamental form is given by

$$\begin{pmatrix} U(u) + V(v) & 0 \\ 0 & U(u) + V(v) \end{pmatrix},$$

where $U(u)$ and $V(v)$ are smooth functions.

Let S be a Liouville surface. Then

$$E(u,v) = U(u) + V(v),$$
$$F(u,v) = 0,$$
$$G(u,v) = U(u) + V(v).$$

Exercise 4.22. Find the Christoffel coefficients of a Liouville surface.

Answer 4.14.

$$\Gamma_{11}^1 = \frac{U'(u)}{2(U(u)+V(v))},$$
$$\Gamma_{11}^2 = -\frac{V'(v)}{2(U(u)+V(v))},$$
$$\Gamma_{12}^1 = \frac{V'(v)}{2(U(u)+V(v))},$$
$$\Gamma_{12}^2 = \frac{U'(u)}{2(U(u)+V(v))},$$
$$\Gamma_{22}^1 = -\frac{U'(u)}{2(U(u)+V(v))},$$
$$\Gamma_{22}^2 = \frac{V'(v)}{2(U(u)+V(v))},$$

where the prime denotes the derivative with respect to the related variable.

Exercise 4.23. Let S be a Liouville surface. Find the equations of the geodesics.

Answer 4.15.

$$2u''(U(u) + V(v)) + (u')^2 U'(u) - 2u'v'V'(v) - (v')^2 U'(u) = 0,$$
$$2v''(U(u) + V(v)) - (u')^2 V'(v) + 2u'v'U'(u) - (v')^2 U'(u) = 0.$$

4.7 Ruled surfaces

Let $I \subseteq \mathbb{R}$, and let S be an oriented surface with a local parameterization $(I \times \mathbb{R}, f)$ that is compatible with the orientation of S.

Definition 4.10. The surface S is called a ruled surface if its local parameterization has the form

$$f(u, v) = g(u) + vb(u), \quad (u, v) \in I \times \mathbb{R},$$

where $g, b \in C^2(I)$ and $|b(u)| = 1, u \in I$.

Exercise 4.24. Let S be a ruled surface. Find the matrix of its first fundamental form.

Solution. We have

$$f_u(u, v) = g'(u) + vb'(u),$$
$$f_v(u, v) = b(u).$$

The components of the first fundamental form are:

$$E = g_{11} = g' \cdot g' + (2v)g' \cdot b' + (v^2)b' \cdot b',$$
$$F = g_{12} = g' \cdot b,$$
$$G = g_{22} = 1.$$

Hence, the sought matrix is

$$\begin{pmatrix} g' \cdot g' + (2v)g' \cdot b' + (v^2)b' \cdot b' & g' \cdot b \\ g' \cdot b & 1 \end{pmatrix}.$$

Exercise 4.25. Let S be a ruled surface. Find the equation of its tangent plane at an arbitrary point.

Solution. Using the calculations of the previous solution, we have

$$f_u(u, v) \times f_v(u, v) = g'(u) \times b(u) + vb'(u) \times b(u).$$

Let R be some point in the tangent plane of S at an arbitrary point. Then, its equation is

$$(R - g - vb) \cdot (g' \times b + vb' \times b) = 0,$$

or

$$(R - g) \cdot (g' \times b + vb' \times b) = 0.$$

Exercise 4.26. Let S be a ruled surface. Find the coefficients L, M, N of the second fundamental form.

Answer 4.16.

$$L = \frac{1}{W}((g' + vb') \times b) \cdot (g'' + vb''),$$
$$M = \frac{1}{W}((g' \times b) \cdot b'),$$
$$N = 0,$$

where $W = \sqrt{EG - F^2}$.

Exercise 4.27. Let S be a ruled surface. Find its Gauss curvature.

Answer 4.17.

$$K = -\frac{((g' \times b) \cdot b')^2}{W^4}.$$

4.8 Minimal surfaces

Let S be an oriented surface with a local parameterization (U, f) that is compatible with the orientation of S.

Definition 4.11. The surface S is said to be minimal if its mean curvature $H = 0$.

In other words, the surface S is minimal if and only if

$$GL - 2FM + EN = 0.$$

Definition 4.12. A local parameterization (U, f) of S is said to be isothermic if

$$E = G, \quad F = 0.$$

Note that the expression $F = 0$ describes that the coordinate lines are orthogonal.

Exercise 4.28. Prove that a surface S is minimal if and only if its asymptotic directions are orthogonal.

Solution. We choose an isothermic representation of S. Then

$$E = G = \lambda^2, \quad F = 0.$$

Hence, the minimality condition is equivalent to the condition

$$E(L + N) = 0,$$

whereupon

$$L + N = 0.$$

The equation of the asymptotic directions is

$$Lt^2 + 2Nt + M = 0$$

and the condition that they are orthogonal is

$$t_1 t_2 = -1,$$

whereupon

$$L = -N,$$

or

$$L + N = 0.$$

This completes the solution.

4.9 Advanced practical problems

Problem 4.1. Find the Christoffel coefficients of the following surfaces:
1.

$$f(t_1, t_2) = (a \cos t_1 \cos t_2, a \cos t_1 \sin t_2, c \sin t_1), \quad (t_1, t_2) \in [0, 2\pi] \times [0, 2\pi],$$

where $a, c \in \mathbb{R}$;
2.

$$f(t_1, t_2) = (t_1 \cos t_2, t_1 \sin t_2, t_1^2), \quad t_1 \in \mathbb{R}, \quad t_2 \in [0, 2\pi];$$

3.

$$f(t_1, t_2) = (R \cos t_2, R \sin t_2, t_1), \quad t_1 \in \mathbb{R}, \quad t_2 \in [0, 2\pi],$$

where $R \in \mathbb{R}$;
4.

$$f(t_1, t_2) = (t_1 \cos t_2, t_1 \sin t_2, kt_1), \quad t_1 \in \mathbb{R}, \quad t_2 \in [0, 2\pi],$$

where $k \in \mathbb{R}$;

5.
$$f(t_1, t_2) = (t_1 \cos t_2, t_1 \sin t_2, at_2), \quad t_1 \in \mathbb{R}, \quad v \in [0, 2\pi],$$

where $a \in \mathbb{R}$.

Answer 4.18. 1.
$$\Gamma^1_{11}(t_1, t_2) = \frac{(a^2 - c^2) \sin(2t_1)}{2(a^2(\sin t_1)^2 + c^2(\cos t_1)^2)},$$
$$\Gamma^2_{11}(t_1, t_2) = 0,$$
$$\Gamma^1_{12}(t_1, t_2) = 0,$$
$$\Gamma^2_{12}(t_1, t_2) = -\tan t_1,$$
$$\Gamma^1_{22}(t_1, t_2) = -\tan t_1,$$
$$\Gamma^2_{22}(t_1, t_2) = 0, \quad (t_1, t_2) \in [0, 2\pi] \times [0, 2\pi];$$

2.
$$\Gamma^1_{11}(t_1, t_2) = \frac{4t_1}{1 + 4t_1^2},$$
$$\Gamma^2_{11}(t_1, t_2) = 0,$$
$$\Gamma^1_{12}(t_1, t_2) = 0,$$
$$\Gamma^2_{12}(t_1, t_2) = \frac{1}{t_1},$$
$$\Gamma^1_{22}(t_1, t_2) = -\frac{t_1}{1 + 4t_1^2},$$
$$\Gamma^2_{22}(t_1, t_2) = 0, \quad t_1 \in \mathbb{R}, \quad t_1 \neq 0, \quad t_2 \in [0, 2\pi];$$

3.
$$\Gamma^1_{11}(t_1, t_2) = 0,$$
$$\Gamma^2_{11}(t_1, t_2) = 0,$$
$$\Gamma^1_{12}(t_1, t_2) = 0,$$
$$\Gamma^2_{12}(t_1, t_2) = 0,$$
$$\Gamma^1_{22}(t_1, t_2) = 0,$$
$$\Gamma^2_{22}(t_1, t_2) = 0, \quad t_1 \in \mathbb{R}, \quad t_2 \in [0, 2\pi];$$

4.
$$\Gamma^1_{11}(t_1, t_2) = 0,$$
$$\Gamma^2_{11}(t_1, t_2) = 0,$$
$$\Gamma^1_{12}(t_1, t_2) = 0,$$
$$\Gamma^2_{12}(t_1, t_2) = \frac{1}{t_1},$$
$$\Gamma^1_{22}(t_1, t_2) = -\frac{t_1}{1 + k^2},$$

$$\Gamma_{22}^2(t_1, t_2) = 0, \quad t_1 \in \mathbb{R}, \quad t_2 \in [0, 2\pi];$$

5.

$$\Gamma_{11}^1(t_1, t_2) = 0,$$

$$\Gamma_{11}^2(t_1, t_2) = 0,$$

$$\Gamma_{12}^1(t_1, t_2) = 0,$$

$$\Gamma_{12}^2(t_1, t_2) = \frac{t_1}{a^2 + t_1^2}$$

$$\Gamma_{22}^1(t_1, t_2) = -t_1,$$

$$\Gamma_{22}^2(t_1, t_2) = 0, \quad t_1 \in \mathbb{R}, \quad t_2 \in [0, 2\pi].$$

Problem 4.2. Let $U \subset \mathbb{R}^2$ and $f \in C^2(U)$. Find the Christoffel coefficients for the surface

$$z = f(x_1, x_2), \quad (x_1, x_2) \in U.$$

Answer 4.19.

$$\Gamma_{ij}^k(x_1, x_2) = \frac{f_{x_i x_j}(x_1, x_2) f_{x_k}(x_1, x_2)}{1 + (f_{x_1}(x_1, x_2))^2 + (f_{x_2}(x_1, x_2))^2}, \quad (x_1, x_2) \in U,$$

with $i, j, k \in \{1, 2\}$.

Problem 4.3 (Gauss theorem). Prove that Gauss curvature of a surface of class at least C^3 depends only on the coefficients of the first fundamental form of the surface and their derivatives up to second order.

Hint 4.20. Start with the formula

$$K = \frac{LN - M^2}{EG - F^2}$$

and rewrite it in the form

$$K(EG - F^2) = LN - M^2.$$

Then, use the representation

$$K(EG - F^2) = ((f_{t_1 t_1} \times f_{t_1}) \cdot f_{t_2})((f_{t_2 t_2} \times f_{t_1}) \cdot f_{t_2}) - ((f_{t_1 t_2} \times f_{t_1}) \cdot f_{t_2})^2.$$

After this, use the well-known formula

$$((a \times b) \cdot c)((d \times e) \cdot f) = \begin{vmatrix} a \cdot d & a \cdot e & a \cdot f \\ b \cdot d & b \cdot e & b \cdot f \\ c \cdot d & c \cdot e & c \cdot f \end{vmatrix}$$

and the relations

$$f_{t_1t_1} \cdot f_{t_1} = \frac{1}{2}E_{t_1},$$

$$f_{t_1t_2} \cdot f_{t_1} = \frac{1}{2}E_{t_2},$$

$$f_{t_2t_2} \cdot f_{t_2} = \frac{1}{2}G_{t_2},$$

$$f_{t_1t_2} \cdot f_{t_2} = \frac{1}{2}G_{t_1},$$

$$f_{t_1t_1} \cdot f_{t_2} = F_{t_1} - \frac{1}{2}E_{t_2},$$

$$f_{t_2t_2} \cdot f_{t_1} = F_{t_2} - \frac{1}{2}G_{t_1}.$$

Then, differentiate the fourth and fifth equations of the latter system with respect to t_1 and t_2, respectively, and get the system

$$f_{t_1t_1t_2} \cdot f_{t_2} + f_{t_1t_2} \cdot f_{t_1t_2} = \frac{1}{2}G_{t_1t_1},$$

$$f_{t_1t_1t_2} \cdot f_{t_2} + f_{t_1t_1} \cdot f_{t_2t_2} = F_{t_1t_2} - \frac{1}{2}E_{t_2t_2},$$

whereupon

$$f_{t_1t_1} \cdot f_{t_2t_2} - f_{t_1t_2} \cdot f_{t_1t_2} = F_{t_1t_2} - \frac{1}{2}E_{t_2t_2} - \frac{1}{2}G_{t_1t_2}.$$

Combining everything, obtain the representation

$$K = \frac{1}{(EG - F^2)^2} \begin{vmatrix} -\frac{1}{2}G_{t_1t_1} + F_{t_1t_2} - \frac{1}{2}E_{t_2t_2} & \frac{1}{2}E_{t_1} & F_{t_1} - \frac{1}{2}E_{t_2} \\ F_{t_2} - \frac{1}{2}G_{t_1} & E & F \\ \frac{1}{2}G_{t_2} & F & G \end{vmatrix}$$

$$- \frac{1}{(EG - F^2)^2} \begin{vmatrix} 0 & \frac{1}{2}E_{t_2} & \frac{1}{2}G_{t_1} \\ \frac{1}{2}E_{t_2} & E & F \\ \frac{1}{2}G_{t_1} & F & G \end{vmatrix}.$$

Problem 4.4 (Frobenius theorem). Prove that Gauss curvature of a surface can be written in the form

$$K_t = -\frac{1}{4(EG - F^2)^2} \begin{vmatrix} E & E_{t_1} & E_{t_2} \\ F & F_{t_1} & F_{t_2} \\ G & G_{t_1} & G_{t_2} \end{vmatrix}$$

$$+ \frac{1}{2\sqrt{EG - F^2}} \left(\frac{\partial}{\partial t_1} \left(\frac{F_{t_2} - G_{t_1}}{\sqrt{EG - F^2}} \right) + \frac{\partial}{\partial t_2} \left(\frac{F_{t_1} - F_{t_2}}{\sqrt{EG - F^2}} \right) \right).$$

Hint 4.21. Use Problem 4.3.

Problem 4.5 (Liouville theorem). Prove that Gauss curvature of a surface can be represented in the form

$$K = -\frac{1}{2\sqrt{EG - F^2}}\left(\frac{\partial}{\partial t_1}\left(\frac{G_{t_1} + \frac{F}{G}G_{t_2} - 2E_{t_1}}{\sqrt{EG - F^2}}\right) + \frac{\partial}{\partial t_2}\left(\frac{E_{t_2} - \frac{F}{G}G_{t_1}}{\sqrt{EG - F^2}}\right)\right).$$

Hint 4.22. Use Problems 4.3 and 4.4.

Problem 4.6. Find the geodesic curvature of the circle with radius $r < R$ on a sphere with radius R.

Answer 4.23.

$$\frac{\sqrt{R^2 - r^2}}{R - r}.$$

Problem 4.7. Find the geodesic curvature of the line $t_1 = $ const. on the surface

$$(t_1 \cos t_2, t_1 \sin t_2, a t_2), \quad t_1 \in \mathbb{R}, \quad t_2 \in [0, 2\pi],$$

where $a > 0$.

Answer 4.24.

$$\frac{|t_1|}{t_1^2 + a^2}.$$

5 Differential forms

In this chapter we first give the differential forms and the operations among them. Exterior differentiation is introduced as the derivations of differential forms. In view of exterior differentiation the exact and closed differential forms are studied. Finally, we concentrate on the well-known operators such as gradient, curl, and divergence operators.

5.1 Algebra of differential forms

We will be interested in $\mathbb{R}^3$ for convenience. Let $\{x, y, z\}$ be canonical coordinates in $\mathbb{R}^3$. At the end of this chapter, we will briefly indicate what happens in $\mathbb{R}^n$.

Definition 5.1. 1. A 0-differential form is a smooth function $f : \mathbb{R}^3 \to \mathbb{R}$.
2. A 1-differential form is

$$\phi = fdx + gdy + hdz,$$

where $f, g, h : \mathbb{R}^3 \to \mathbb{R}$ are given smooth functions.
3. A 2-differential form is

$$\phi = fdxdy + gdydz + hdxdz,$$

where $f, g, h : \mathbb{R}^3 \to \mathbb{R}$ are given smooth functions.
4. A 3-differential form is

$$\phi = fdxdydz,$$

where $f : \mathbb{R}^3 \to \mathbb{R}$ is a given smooth function.

Example 5.1. The form

$$\phi = x + y + z$$

is a 0-differential form.

Example 5.2. The form

$$\phi = xdx + ydy + zdz$$

is a 1-differential form.

Example 5.3. The form

$$\phi = dydx + dxdz$$

is a 2-differential form.

https://doi.org/10.1515/9783111501857-005

Example 5.4. The form

$$\phi = \sin(x + y + z)dxdydz$$

is a 3-differential form.

Definition 5.2. Let

$$\phi_1 = f_1 dx + g_1 dy + h_1 dz,$$
$$\phi_2 = f_2 dx + g_2 dy + h_2 dz$$

be two 1-differential forms. Then we define

$$\phi_1 + \phi_2 = (f_1 + f_2)dx + (g_1 + g_2)dy + (h_1 + h_2)dz$$

and

$$\phi_1 - \phi_2 = (f_1 - f_2)dx + (g_1 - g_2)dy + (h_1 - h_2)dz.$$

Definition 5.3. Let

$$\phi_1 = f_1 dxdy + g_1 dydz + h_1 dxdz,$$
$$\phi_2 = f_2 dxdy + g_2 dydz + h_2 dxdz$$

be given 2-differential forms. Then we define

$$\phi_1 + \phi_2 = (f_1 + f_2)dxdy + (g_1 + g_2)dydz + (h_1 + h_2)dxdz$$

and

$$\phi_1 - \phi_2 = (f_1 - f_2)dxdy + (g_1 - g_2)dydz + (h_1 - h_2)dxdz.$$

Definition 5.4. Let

$$\phi_1 = f_1 dxdydz,$$
$$\phi_2 = f_2 dxdydz$$

be two 3-differential forms. Then we define

$$\phi_1 + \phi_2 = (f_1 + f_2)dxdydz$$

and

$$\phi_1 - \phi_2 = (f_1 - f_2)dxdydz.$$

Remark 5.1. Note that the addition and subtraction make sense only for k-differential forms, $k \in \{0,1,2,3\}$, and not for a k-differential form and for an l-differential form, $k \neq l, k, l \in \{0,1,2,3\}$.

Example 5.5. Let

$$\phi = (x^2 + 2xy)dx + (xy - 1)dy + x^3 dz,$$
$$\psi = (x^2 - 2xy)dx + dy + x^3 dz.$$

Then

$$\phi + \psi = 2x^2 dx + xydy + 2x^3 dz$$

and

$$\phi - \psi = 4xydx + (xy - 2)dy.$$

Exercise 5.1. Let

$$\phi = (x - y + 3y^2)dxdy - dxdz,$$
$$\psi = (x^2 + y - 2y^2)dxdy + (x - 3x^2 y)dydz + 3dxdz.$$

Find
1. $\phi + \psi$;
2. $\phi - \psi$.

Answer 5.1. 1.
$$(x^2 + x + y^2)dxdy + (x - 3x^2 y)dydz + 2dxdz;$$

2.
$$(x - x^2 - 2y + 5y^2)dxdy + (3x^2 y - x)dydz - dxdz.$$

Definition 5.5. Let $m, f, g, h : \mathbb{R}^3 \to \mathbb{R}$ and

$$\phi = fdx + gdy + hdz.$$

Then, we define $m\phi$ as follows:

$$m\phi = mfdx + mgdy + mhdz.$$

Definition 5.6. Let $m, f, g, h : \mathbb{R}^3 \to \mathbb{R}$ and

$$\phi = fdxdy + gdydz + hdxdz.$$

Then, we define $m\phi$ as follows:

$$m\phi = mfdxdy + mgdydz + mhdxdz.$$

Definition 5.7. Let $m, f : \mathbb{R}^3 \to \mathbb{R}$ and

$$\phi = f dx dy dz.$$

Then, we define $m\phi$ as follows:

$$m\phi = mf dx dy dz.$$

Example 5.6. Let

$$\phi = 2x^3 dxdy + xdydz + x^2 y^4 dxdz.$$

Then

$$3\phi = 6x^3 dxdy + 3xdydz + 3x^2 y^4 dxdz.$$

Next,

$$\frac{1}{x}\phi = 2x^2 dxdy + dydz + xy^4 dxdz.$$

Exercise 5.2. Let

$$\phi_1 = 2x^2 dx + (x + y)dy,$$
$$\phi_2 = -xdx + (x - 2y)dy.$$

Find
1. $2\phi_1 + \phi_2$;
2. $\phi_1 - x\phi_2$.

Answer 5.2. 1.
$$(4x^2 - x)dx + 3xdy;$$

2.
$$3x^2 dx + (-x^2 + 2xy + x + y)dy.$$

For the multiplication of differential forms, we introduce the following rules:

$$dxdx = 0,$$
$$dydy = 0,$$
$$dzdz = 0,$$

and

$$dxdy = -dydx,$$

$$dxdz = -dzdx,$$
$$dydz = -dzdy.$$

Example 5.7. We have

$$dxdydx + dxdy + 2dydx = -dxdxdy + dxdy - 2dxdy$$
$$= dxdy - 2dxdy$$
$$= -dxdy.$$

Example 5.8. We have

$$dx(dx + dy + dz) = dxdx + dxdy + dxdz$$
$$= dxdy + dxdz$$
$$= dxdy - dzdx.$$

Example 5.9. Let

$$\phi = x^3 ydx + ydy,$$
$$\psi = x^4 dx + xdy + z^2 dz,$$

and

$$\rho = xyzdzdx.$$

Then

$$\phi\psi = (x^3 ydx + ydy)(x^4 dx + xdy + z^2 dz)$$
$$= x^7 ydxdx + x^4 ydydx + x^4 ydxdy + yxdydy + x^3 yz^2 dxdz + z^2 ydydz$$
$$= -x^4 ydxdy + x^4 ydxdy + x^3 yz^2 dxdz + z^2 ydydz$$
$$= x^3 yz^2 dxdz + z^2 ydydz.$$

Next,

$$\phi\rho = (x^3 ydx + ydy)xyzdzdx$$
$$= x^4 y^2 zdxdzdx + xy^2 zdydzdx$$
$$= xy^2 zdxdydz.$$

Exercise 5.3. Let

$$\phi_1 = 2x^2 dx + (x + y)dy,$$
$$\phi_2 = -xdx + (x - 2y)dy,$$

$$\phi_3 = x^3 dx + yzdy - (x^2 + y^2 + z^2)dz,$$
$$\phi_4 = y^2 zdx - xzdy + (2x + 1)dz.$$

Find

1. $\phi_1\phi_2$;
2. $\phi_1\phi_3$;
3. $\phi_1\phi_4$;
4. $\phi_2\phi_3$;
5. $\phi_2\phi_4$;
6. $\phi_3\phi_4$.

Answer 5.3. 1.

$$(2x^3 - 4x^2 y + x^2 + xy)dxdy;$$

2.

$$-(x + y)(x^2 + y^2 + z^2)dydz + 2x^2(x^2 + y^2 + z^2)dzdx + (2x^2 yz - x^4 - x^3 y)dxdy;$$

3.

$$(x + y)(2x + 1)dydz - 2x^2(2x + 1)dzdx - (2x^3 z + xy^2 z + y^3 z)dxdy;$$

4.

$$-(x - 2y)(x^2 + y^2 + z^2)dydz - x(x^2 + y^2 + z^2)dzdx - (xyz + x^4 - 2x^3 y)dxdy;$$

5.

$$(x - 2y)(2x + 1)dydz + x(2x + 1)dzdx + (x^2 z - xy^2 z + 2y^3 z)dxdy;$$

6.

$$(yz(2x + 1) - xz(x^2 + y^2 + z^2))dydz - (y^2 z(x^2 + y^2 + z^2) + 2x^4 + x^3)dzdx$$
$$- (x^4 z + y^3 z^2)dxdy.$$

Example 5.10. Let ϕ be a k-differential form. We will prove that

$$\phi^2 = 0.$$

For this, we will consider the following cases:

1. Let

$$\phi = fdx + gdy + hdz,$$

where $f, g, h : \mathbb{R}^3 \rightarrow \mathbb{R}$. Then

$$\phi^2 = (fdx + gdy + hdz)(fdx + gdy + hdz)$$
$$= f^2 dxdx + fgdxdy + fhdxdz + gfdydx + g^2 dydy + ghdydz$$
$$+ hfdzdx + hgdzdy + h^2 dzdz$$
$$= fgdxdy - fgdxdy + fhdxdz - fhdxdz + ghdydz - ghdydz$$
$$= 0.$$

2. Let

$$\phi = f dx dy + g dy dz + h dx dz,$$

where $f, g, h : \mathbb{R}^3 \to \mathbb{R}$. Then

$$
\begin{aligned}
\phi^2 &= (f dx dy + g dy dz + h dx dz)(f dx dy + g dy dz + h dx dz) \\
&= f^2 dx dy dx dy + f g dx dy dy dz + f h dx dz dx dz + g f dy dz dx dy \\
&\quad + g^2 dy dz dy dz + g h dx dz dx dz \\
&\quad + h f dx dz dx dy + h g dx dz dy dz + h^2 dx dz dx dz \\
&= 0.
\end{aligned}
$$

3. Let

$$\phi = f dx dy dz,$$

where $f : \mathbb{R}^3 \to \mathbb{R}$. Then

$$
\begin{aligned}
\phi^2 &= \phi \phi \\
&= (f dx dy dz)(f dx dy dz) \\
&= f^2 dx dy dz dx dy dz \\
&= 0.
\end{aligned}
$$

This completes the solution.

Example 5.11. Let

$$
\begin{aligned}
\phi_1 &= f_1 dx + g_1 dy + h_1 dz, \\
\phi_2 &= f_2 dx + g_2 dy + h_2 dz.
\end{aligned}
$$

We will show that

$$\phi_1 \phi_2 = (f_1 g_2 - g_1 f_2) dx dy + (f_1 h_2 - h_1 f_2) dx dz + (g_1 h_2 - h_1 g_2) dy dz.$$

Really, we have

$$
\begin{aligned}
\phi_1 \phi_2 &= (f_1 dx + g_1 dy + h_1 dz)(f_2 dx + g_2 dy + h_2 dz) \\
&= f_1 f_2 dx dx + f_1 g_2 dx dy + f_1 h_2 dx dz + g_1 f_2 dy dx + g_1 g_2 dy dy + g_1 h_2 dy dz \\
&\quad + h_1 f_2 dz dx + h_1 g_2 dz dy + h_1 h_2 dz dz \\
&= (f_1 g_2 - g_1 f_2) dx dy + (f_1 h_2 - h_1 f_2) dx dz + (g_1 h_2 - h_1 g_2) dz dy.
\end{aligned}
$$

Example 5.12. Let

$$\phi_1 = f_1 dx + g_1 dy + h_1 dz,$$
$$\phi_2 = f_2 dxdy + g_2 dydz + h_2 dzdx.$$

We will prove that

$$\phi_1\phi_2 = (f_1 g_2 + g_1 h_2 + h_1 f_2)dxdydz.$$

Really, we have

$$\phi_1\phi_2 = (f_1 dx + g_1 dy + h_1 dz)(f_2 dxdy + g_2 dydz + h_2 dzdx)$$
$$= f_1 f_2 dxdxdy + f_1 g_2 dxdydz + f_1 h_2 dxdzdx$$
$$+ g_1 f_2 dydxdy + g_1 g_2 dydydz + g_1 h_2 dydzdx$$
$$+ h_1 f_2 dzdxdy + h_1 g_2 dzdydz + h_1 h_2 dzdzdx$$
$$= (f_1 g_2 + g_1 h_2 + h_1 f_2)dxdydz.$$

5.2 Exterior differentiation

In this section, we introduce exterior differentiation of differential forms.

Definition 5.8. Let f be a 0-differential form. Then its exterior derivative is defined by

$$df = f_x dx + f_y dy + f_z dz.$$

Example 5.13. Let

$$f(x,y,z) = \frac{x^2 + y^2}{z}, \quad (x,y,z) \in \mathbb{R}^3.$$

Then

$$f_x(x,y,z) = \frac{2x}{z},$$
$$f_y(x,y,z) = \frac{2y}{z},$$
$$f_z(x,y,z) = -\frac{x^2 + y^2}{z^2}, \quad (x,y,z) \in \mathbb{R}^3.$$

Hence,

$$df = f_x dx + f_y dy + f_z dz$$
$$= \frac{2x}{z} dx + \frac{2y}{z} dy - \frac{x^2 + y^2}{z^2} dz, \quad (x,y,z) \in \mathbb{R}^3.$$

Example 5.14. Let

$$f(x,y,z) = x + y + z, \quad (x,y,z) \in \mathbb{R}^3.$$

Then

$$f_x(x,y,z) = 1,$$
$$f_y(x,y,z) = 1,$$
$$f_z(x,y,z) = 1, \quad (x,y,z) \in \mathbb{R}^3.$$

Therefore,

$$df(x,y,z) = f_x(x,y,z)dx + f_y(x,y,z)dy + f_z(x,y,z)dz$$
$$= dx + dy + dz, \quad (x,y,z) \in \mathbb{R}^3.$$

Exercise 5.4. Let

$$f_1(x,y,z) = x^2 y^3 z - 2xyz^2,$$
$$f_2(x,y,z) = x^2 + y^2 - 3z^4,$$
$$f_3(x,y,z) = x^2 y + y^2 z,$$
$$f_4(x,y,z) = 2x + 3y - 4z + 5, \quad (x,y,z) \in \mathbb{R}^3.$$

Find
1. df_1;
2. df_2;
3. df_3;
4. df_4.

Answer 5.4. 1.

$$(2xy^3 z - 2yz^2)dx + (3x^2 y^2 z - 2xz^2)dy + (x^2 y^3 - 4xyz)dz, \quad (x,y,z) \in \mathbb{R}^3;$$

2.

$$2xdx + 2ydy - 12z^3 dz, \quad (x,y,z) \in \mathbb{R}^3;$$

3.

$$2xydx + (x^2 + 2yz)dy + y^2 dz, \quad (x,y,z) \in \mathbb{R}^3;$$

4.

$$2dx + 3dy - 4dz, \quad (x,y,z) \in \mathbb{R}^3.$$

Definition 5.9. Let ϕ be a k-differential form. Then its exterior derivative $d\phi$ is a $(k+1)$-differential form obtained from ϕ by applying d to each of the functions included in ϕ.

1. Let

$$\phi = f dx + g dy + h dz,$$

where $f, g, h : \mathbb{R}^3 \to \mathbb{R}$ are given functions. Then

$$
\begin{aligned}
d\phi &= (df)dx + (dg)dy + (dh)dz \\
&= (f_x dx + f_y dy + f_z dz)dx + (g_x dx + g_y dy + g_z dz)dy \\
&\quad + (h_x dx + h_y dy + h_z dz)dz \\
&= f_y dy dx + f_z dz dx + g_x dx dy + g_z dz dy + h_x dx dz + h_y dy dz \\
&= (g_x - f_y)dx dy + (h_x - f_z)dx dz + (h_y - g_z)dy dz.
\end{aligned}
$$

Example 5.15. Let

$$\phi = (y^2 + z^3)dx + (x + z^3)dy + (x + y^2)dz, \quad (x, y, z) \in \mathbb{R}^3.$$

We have

$$
\begin{aligned}
f(x, y, z) &= y^2 + z^3, \\
g(x, y, z) &= x + z^3, \\
h(x, y, z) &= x + y^2, \quad (x, y, z) \in \mathbb{R}^3.
\end{aligned}
$$

Hence,

$$
\begin{aligned}
f_y(x, y, z) &= 2y, \\
f_z(x, y, z) &= 3z^2, \\
g_x(x, y, z) &= 1, \\
g_z(x, y, z) &= 3z^2, \\
h_x(x, y, z) &= 1, \\
h_y(x, y, z) &= 2y, \quad (x, y, z) \in \mathbb{R}^3,
\end{aligned}
$$

and

$$d\phi = (1 - 2y)dx dy + (1 - 3z^2)dx dz + (2y - 3z^2)dy dz,$$

for $(x, y, z) \in \mathbb{R}^3$.

Exercise 5.5. Let

$$\phi = (x + y + z)dx + (x^2 + y^2 + z^2)dy + (x + 2y + 3z)dz.$$

Find $d\phi$.

Answer 5.5.

$$(2x - 1)dxdy + 2(1 - z)dydz, \quad (x, y, z) \in \mathbb{R}^3.$$

2. Let

$$\phi = fdxdy + gdydz + hdzdx,$$

where $f, g, h : \mathbb{R}^3 \to \mathbb{R}$. Then

$$
\begin{aligned}
d\phi &= (f_x dx + f_y dy + f_z dz)dxdy \\
&\quad + (g_x dx + g_y dy + g_z dz)dydz \\
&\quad + (h_x dx + h_y dy + h_z dz)dzdx \\
&= f_z dxdydz + g_x dxdydz + h_y dxdydz \\
&= (f_z + g_x + h_y)dxdydz.
\end{aligned}
$$

Example 5.16. Let

$$\phi = (x^2 - y^2 - z^2)dxdy + (x + z)dydz + y^3 dzdx.$$

Here

$$
\begin{aligned}
f(x, y, z) &= x^2 - y^2 - z^2, \\
g(x, y, z) &= x + z, \\
h(x, y, z) &= y^3, \quad (x, y, z) \in \mathbb{R}^3,
\end{aligned}
$$

and

$$
\begin{aligned}
f_z(x, y, z) &= -2z, \\
g_x(x, y, z) &= 1, \\
h_y(x, y, z) &= 3y^2, \quad (x, y, z) \in \mathbb{R}^3.
\end{aligned}
$$

Therefore,

$$d\phi = (-2z + 1 + 3y^2)dxdydz.$$

Exercise 5.6. Let

$$\phi = \frac{x + y + z}{1 + x^2 + y^2 + z^2}(dxdy + dydz + dzdx), \quad (x, y, z) \in \mathbb{R}^3.$$

Find $d\phi$.

Answer 5.6.

$$d\phi = \frac{1 + x^2 + y^2 + z^2 - 4xy - 4xz - 4yz}{(1 + x^2 + y^2 + z^2)^2} dxdydz, \quad (x, y, z) \in \mathbb{R}^3.$$

5.3 Properties of the exterior differentiation

In this section, we will deduce some of the properties of the exterior differentiation of the differential forms:

1. The exterior differentiation is a linear operation.
    For the proof, we will consider the following cases.
    a. Let ϕ_1, ϕ_2 be 0-differential forms and $a_1, a_2 \in \mathbb{R}$. Then

$$d\phi_1 = \phi_{1x}dx + \phi_{1y}dy + \phi_{1z}dz,$$
$$d(a_1\phi_1) = (a_1\phi_{1x})dx + (a_1\phi_{1y})dy + (a_1\phi_{1z})dz$$
$$= a_1(\phi_{1x}dx + \phi_{1y}dy + \phi_{1z}dz)$$
$$= a_1 d\phi_1,$$
$$d\phi_2 = \phi_{2x}dx + \phi_{2y}dy + \phi_{2z}dz,$$
$$d(a_2\phi_2) = (a_2\phi_{2x})dx + (a_2\phi_{2y})dy + (a_2\phi_{2z})dz$$
$$= a_2(\phi_{2x}dx + \phi_{2y}dy + \phi_{2z}dz)$$
$$= a_2 d\phi_2$$

and

$$d(a_1\phi_1 + a_2\phi_2) = (a_1\phi_{1x} + a_2\phi_{2x})dx + (a_1\phi_{1y} + a_2\phi_{2y})dy$$
$$+ (a_1\phi_{1z} + a_2\phi_{2z})dz$$
$$= (a_1\phi_{1x})dx + (a_1\phi_{1y})dy + (a_1\phi_{1z})dz$$
$$+ a_2(\phi_{2x}dx + \phi_{2y}dy + \phi_{2z}dz)$$
$$= a_1 d\phi_1 + a_2 d\phi_2.$$

b. Let ϕ_1 and ϕ_2 be two 1-differential forms, i. e.,

$$\phi_1 = f_1 dx + g_1 dy + h_1 dz,$$
$$\phi_2 = f_2 dx + g_2 dy + h_2 dz,$$

where $f_1, f_2, g_1, g_2, h_1, h_2 : \mathbb{R}^3 \to \mathbb{R}$ are given smooth functions. Let also $a_1, a_2 \in \mathbb{R}$. Then

$$d\phi_1 = (g_{1x} - f_{1y})dxdy + (f_{1z} - h_{1x})dzdx + (h_{1y} - g_{1z})dydz,$$
$$d(a_1\phi_1) = (a_1 g_{1x} - a_1 f_{1y})dxdy + (a_1 f_{1z} - a_1 h_{1x})dzdx + (a_1 h_{1y} - a_1 g_{1z})dydz$$

$$= a_1((g_{1x} - f_{1y})dxdy + (f_{1z} - h_{1x})dzdx + (h_{1y} - g_{1z})dydz)$$
$$= a_1 d\phi_1,$$

and

$$d\phi_2 = (g_{2x} - f_{2y})dxdy + (f_{2z} - h_{2x})dzdx + (h_{2y} - g_{2z})dydz,$$
$$d(a_2\phi_2) = (a_2 g_{2x} - a_2 f_{2y})dxdy + (a_2 f_{2z} - a_2 h_{2x})dzdx$$
$$+ (a_2 h_{2y} - a_2 g_{2z})dydz$$
$$= a_2((g_{2x} - f_{2y})dxdy + (f_{2z} - h_{2x})dzdx + (h_{2y} - g_{2z})dydz)$$
$$= a_2 d\phi_2.$$

Consequently,

$$a_1\phi_1 + a_2\phi_2 = (a_1 f_1 + a_2 f_2)dx + (a_1 g_1 + a_2 g_2)dy + (a_1 h_1 + a_2 h_2)dz$$

and

$$d(a_1\phi_1 + a_2\phi_2) = (a_1 g_{1x} + a_2 g_{2x} - a_1 f_{1y} - a_2 f_{2y})dxdy$$
$$+ (a_1 f_{1z} + a_2 f_{2z} - a_1 h_{1x} - a_2 h_{2x})dzdx$$
$$+ (a_1 h_{1y} + a_2 h_{2y} - a_1 g_{1z} - a_2 g_{2z})dydz$$
$$= a_1((g_{1x} - f_{1y})dxdy + (f_{1z} - h_{1x})dzdx + (h_{1y} - g_{1z})dydz)$$
$$+ a_2((g_{2x} - f_{2y})dxdy + (f_{2z} - h_{2x})dzdx$$
$$+ (h_{2y} - g_{2z})dydz)$$
$$= a_1 d\phi_1 + a_2 d\phi_2.$$

c. Let ϕ_1 and ϕ_2 be two 2-differential forms, i. e.,

$$\phi_1 = f_1 dxdy + g_1 dydz + h_1 dzdy,$$
$$\phi_2 = f_2 dxdy + g_2 dydz + h_2 dzdy,$$

where $f_1, f_2, g_1, g_2, h_1, h_2 : \mathbb{R}^3 \to \mathbb{R}$ are given functions. Let also $a_1, a_2 \in \mathbb{R}$. Then

$$d\phi_1 = (f_{1z} + g_{1x} + h_{1y})dxdydz,$$
$$d(a_1\phi_1) = (a_1 f_{1z} + a_1 g_{1x} + a_1 h_{1y})dxdydz$$
$$= a_1(f_{1z} + g_{1x} + h_{1y})dxdydz$$
$$= a_1 d\phi_1,$$

and

$$d\phi_2 = (f_{2z} + g_{2x} + h_{2y})dxdydz,$$

$$d(a_2\phi_2) = (a_2f_{2z} + a_2g_{2x} + a_2h_{2y})dxdydz$$
$$= a_2(f_{2z} + g_{2x} + h_{2y})dxdydz$$
$$= a_2d\phi_2.$$

Consequently,

$$a_1\phi_1 + a_2\phi_2 = a_1f_1dxdy + a_1g_1dydz + a_1h_1dzdx$$
$$+ a_2f_2dxdy + a_2g_2dydz + a_2h_2dzdx$$
$$= (a_1f_1 + a_2f_2)dxdy + (a_1g_1 + a_2g_2)dydz$$
$$+ (a_1h_1 + a_2h_2)dzdx$$

and

$$d(a_1\phi_1 + a_2\phi_2) = (a_1f_{1z} + a_2f_{2z} + a_1g_{1x} + a_2g_{2x} + a_1h_{1y} + a_2h_{2y})dxdydz$$
$$= a_1(f_{1z} + g_{1x} + h_{1y})dxdydz + a_2(f_{2z} + g_{2x} + h_{2y})dxdydz$$
$$= a_1d\phi_1 + a_2d\phi_2.$$

This completes the proof.

2. Let ϕ_1 and ϕ_2 be 0-differential forms. Then

$$d(\phi_1\phi_2) = \phi_1d\phi_2 + \phi_2d\phi_1.$$

For the proof, we have

$$d\phi_1 = \phi_{1x}dx + \phi_{1y}dy + \phi_{1z}dz,$$
$$d\phi_2 = \phi_{2x}dx + \phi_{2y}dy + \phi_{2z}dz,$$

and

$$d(\phi_1\phi_2) = (\phi_1\phi_2)_xdx + (\phi_1\phi_2)_ydy + (\phi_1\phi_2)_zdz$$
$$= (\phi_1\phi_{2x} + \phi_{1x}\phi_2)dx + (\phi_1\phi_{2y} + \phi_{1y}\phi_2)dy + (\phi_1\phi_{2z} + \phi_{1z}\phi_2)dz$$
$$= \phi_1(\phi_{2x}dx + \phi_{2y}dy + \phi_{2z}dz) + \phi_2(\phi_{1x}dx + \phi_{1y}dy + \phi_{1z}dz)$$
$$= \phi_1d\phi_2 + \phi_2d\phi_1.$$

This completes the proof.

3. Let ϕ_1 be a 0-differential form and ϕ_2 be a 1-form,

$$\phi_2 = fdx + gdy + hdz.$$

Then

$$d(\phi_1\phi_2) = \phi_1d\phi_2 + \phi_2d\phi_1.$$

For the proof, we have

$$d\phi_2 = (g_x - f_y)dxdy + (f_z - h_x)dzdx + (h_y - g_z)dydz$$

and

$$\phi_1\phi_2 = (\phi_1 f)dx + (\phi_1 g)dy + (\phi_1 h)dz.$$

Hence,

$$
\begin{aligned}
d(\phi_1\phi_2) &= (\phi_{1x}g + \phi_1 g_x - \phi_1 f_y - \phi_{1y}f)dxdy \\
&\quad + (\phi_{1z}f + \phi_1 f_z - \phi_{1x}h - \phi_1 h_x)dzdx \\
&\quad + (\phi_{1y}h + \phi_1 h_y - \phi_{1z}g - \phi_1 g_z)dydz \\
&= \phi_1(g_x - f_y)dxdy + \phi_1(f_z - h_x)dzdx \\
&\quad + \phi_1(h_y - g_z)dydz + (\phi_{1z}f - \phi_{1x}h)dzdx \\
&\quad + (\phi_{1x}g - \phi_{1y}f)dxdy + (\phi_{1y}h - \phi_{1z}g)dydz \\
&= \phi_1 d\phi_2 + (\phi_{1x}g - \phi_{1y}f)dxdy \\
&\quad + (\phi_{1z}f - \phi_{1x}h)dzdx + (\phi_{1y}h - \phi_{1z}g)dydz.
\end{aligned}
$$

Next,

$$d\phi_1 = \phi_{1x}dx + \phi_{1y}dy + \phi_{1z}dz$$

and

$$
\begin{aligned}
d\phi_1\phi_2 &= (\phi_{1x}dx + \phi_{1y}dy + \phi_{1z}dz)(fdx + gdy + hdz) \\
&= \phi_{1x}fdxdx + \phi_{1x}gdxdy + \phi_{1x}hdxdz + \phi_{1y}fdydx \\
&\quad + \phi_{1y}gdydy + \phi_{1y}hdydz + \phi_{1z}fdzdx + \phi_{1z}gdzdy + \phi_{1z}hdzdz \\
&= (\phi_{1z}f - \phi_{1x}h)dzdx + (\phi_{1y}h - \phi_{1z}g)dydz + (\phi_{1x}g - \phi_{1y}f)dxdy.
\end{aligned}
$$

Consequently,

$$d(\phi_1\phi_2) = d\phi_1\phi_2 + \phi_1 d\phi_2.$$

This completes the proof.

4. Let ϕ_1 be a 0-differential form and

$$\phi_2 = fdxdy + gdydz + hdzdx.$$

Then

$$d(\phi_1\phi_2) = d\phi_1\phi_2 + \phi_1 d\phi_2.$$

For the proof, we have

$$d\phi_2 = (f_z + g_x + h_y)dxdydz$$

and

$$\phi_1\phi_2 = (\phi_1 f)dxdy + (\phi_1 g)dydz + (\phi_1 h)dzdx.$$

Hence,

$$
\begin{aligned}
d(\phi_1\phi_2) &= ((\phi_1 f)_z + (\phi_1 g)_x + (\phi_1 h)_y)dxdydz \\
&= (\phi_{1z}f + \phi_1 f_z + \phi_{1x}g + \phi_1 g_x + \phi_{1y}h + \phi_1 h_y)dxdydz \\
&= \phi_1(f_z + g_x + h_y)dxdydz + (\phi_{1z}f + \phi_{1x}g + \phi_{1y}h)dxdydz \\
&= \phi_1 d\phi_2 + (\phi_{1z}f + \phi_{1x}g + \phi_{1y}h)dxdydz.
\end{aligned}
$$

Also,

$$
\begin{aligned}
d\phi_1\phi_2 &= (\phi_{1x}dx + \phi_{1y}dy + \phi_{1z}dz)(fdxdy + gdydz + hdzdx) \\
&= \phi_{1x}fdxdxdy + \phi_{1x}gdxdydz + \phi_{1x}hdxdzdx \\
&\quad + \phi_{1y}fdydxdy + \phi_{1y}gdydydz + \phi_{1y}hdydzdx \\
&\quad + \phi_{1z}fdzdxdy + \phi_{1z}gdzdydz + \phi_{1z}hdzdzdx \\
&= (\phi_{1x}g + \phi_{1y}h + \phi_{1z}f)dxdydz.
\end{aligned}
$$

Consequently,

$$d(\phi_1\phi_2) = d\phi_1\phi_2 + \phi_1 d\phi_2.$$

This completes the proof.

5. Let

$$
\begin{aligned}
\phi_1 &= f_1 dx + g_1 dy + h_1 dz, \\
\phi_2 &= f_2 dx + g_2 dy + h_2 dz.
\end{aligned}
$$

Then

$$d(\phi_1\phi_2) = d\phi_1\phi_2 - \phi_1 d\phi_2.$$

For the proof, we have

$$\phi_1\phi_2 = (f_1 g_2 - g_1 f_2)dxdy + (h_1 f_2 - f_1 h_2)dzdx + (g_1 h_2 - h_1 g_2)dydz.$$

Hence,

$$d(\phi_1\phi_2) = ((f_1g_2 - g_1f_2)_z + (h_1f_2 - f_1h_2)_y - (g_1h_2 - h_1g_2)_x)dxdydz$$
$$= (f_{1z}g_2 + f_1g_{2z} - g_{1z}f_2 - g_1f_{2z} + h_{1y}f_2 + h_1f_{2y} - f_{1y}h_2 - f_1h_{2y}$$
$$+ g_{1x}h_2 + g_1h_{2x} - h_{1x}g_2 - h_1g_{2x})dxdydz$$

and

$$d\phi_1 = (g_{1x} - f_{1y})dxdy + (f_{1z} - h_{1x})dzdx + (h_{1y} - g_{1z})dydz,$$

and

$$d\phi_1\phi_2 = ((g_{1x} - f_{1y})dxdy + (f_{1z} - h_{1x})dzdx + (h_{1y} - g_{1z})dydz)$$
$$\cdot (f_2dx + g_2dy + h_2dz)$$
$$= f_2(h_{1y} - g_{1z})dydzdx + g_2(f_{1z} - h_{1x})dzdxdy + h_2(g_{1x} - f_{1y})dxdydz$$
$$= (f_2(h_{1y} - g_{1z}) + g_2(f_{1z} - h_{1x}) + h_2(g_{1x} - f_{1y}))dxdydz.$$

Moreover,

$$d\phi_2 = (g_{2x} - f_{2y})dxdy + (f_{2z} - h_{2x})dzdx + (h_{2y} - g_{2z})dydz,$$

and

$$\phi_1 d\phi_2 = (f_1dx + g_1dy + h_1dz)$$
$$\cdot ((g_{2x} - f_{2y})dxdy + (f_{2z} - h_{2x})dzdx + (h_{2y} - g_{2z})dydz)$$
$$= (f_1(h_{2y} - g_{2z}) + g_1(f_{2z} - h_{2x}) + h_1(g_{2x} - f_{2y}))dxdydz.$$

Therefore,

$$d\phi_1\phi_2 + \phi_1 d\phi_2 = (f_{1z}g_2 + f_1g_{2z} - g_{1z}f_2 - g_1f_{2z}$$
$$+ h_{1y}f_2 + h_1f_{2y} - f_{1y}h_2 - f_1h_{2y}$$
$$+ g_{1x}h_2 + g_1h_{2x} - h_{1x}g_2 - h_1g_{2x})dxdydz$$
$$= d(\phi_1\phi_2).$$

This completes the proof.
6. Let ϕ be a 0-differential form. Then

$$d(d\phi) = 0.$$

For the proof, we have

$$d\phi = \phi_x dx + \phi_y dy + \phi_z dz$$

and

$$d(d\phi) = d(\phi_x dx + \phi_y dy + \phi_z dz)$$
$$= (\phi_{yx} - \phi_{xy})dxdy + (\phi_{xz} - \phi_{zx})dzdx + (\phi_{zy} - \phi_{yz})dydz$$
$$= 0.$$

This completes the proof.

7. Let ϕ be a 1-differential form,

$$\phi = fdx + gdy + hdz.$$

Then

$$d(d\phi) = 0.$$

For the proof, we have

$$d\phi = (g_x - f_y)dxdy + (f_z - h_x)dzdx + (h_y - g_z)dydz.$$

Then

$$d(d\phi) = d((g_x - f_y)dxdy + (f_z - h_x)dzdx + (h_y - g_z)dydz)$$
$$= ((g_x - f_y)_z + (f_z - h_x)_y + (h_y - g_z)_x)dxdydz$$
$$= (g_{xz} - f_{yz} + f_{zy} - h_{xy} + h_{yx} - g_{zx})dxdydz$$
$$= 0.$$

This completes the proof.

5.4 Closed and exact differential forms

Definition 5.10. A differential form ϕ is said to be closed if $d\phi = 0$.

Example 5.17. Let

$$\phi(x,y,z) = 5x^4 y^2 z^3 dx + 2x^5 yz^3 dy + 3x^5 y^2 z^2 dz, \quad (x,y,z) \in \mathbb{R}^3.$$

Here

$$f(x,y,z) = 5x^4 y^2 z^3,$$
$$g(x,y,z) = 2x^5 yz^3,$$
$$h(x,y,z) = 3x^5 y^2 z^2, \quad (x,y,z) \in \mathbb{R}^3.$$

Then

$$f_y(x,y,z) = 10x^4 yz^3,$$

$$f_z(x,y,z) = 15x^4y^2z^2,$$
$$g_x(x,y,z) = 10x^4yz^3,$$
$$g_z(x,y,z) = 6x^5yz^2,$$
$$h_x(x,y,z) = 15x^4y^2z^2,$$
$$h_y(x,y,z) = 6x^5yz^2, \quad (x,y,z) \in \mathbb{R}^3.$$

Hence,

$$d\phi(x,y,z) = (g_x(x,y,z) - f_y(x,y,z))dxdy$$
$$+ (f_z(x,y,z) - h_x(x,y,z))dzdx$$
$$+ (h_y(x,y,z) - g_z(x,y,z))dydz$$
$$= (10x^4yz^3 - 10x^4yz^3)dxdy + (15x^4y^2z^2 - 15x^4y^2z^2)dzdx$$
$$+ (6x^5yz^2 - 6x^5yz^2)dydx$$
$$= 0.$$

Thus, ϕ is a closed differential form on $\mathbb{R}^3$.

Example 5.18. Let

$$\phi(x,y,z) = \left(\frac{x}{z} - 2z\right)dydz + \left(x^2z - \frac{y}{z}\right)dzdx, \quad (x,y,z) \in \mathbb{R}^3.$$

We have

$$g(x,y,z) = \frac{x}{z} - 2z,$$
$$h(x,y,z) = x^2z - \frac{y}{z}, \quad (x,y,z) \in \mathbb{R}^3,$$

and

$$g_x(x,y,z) = \frac{1}{z},$$
$$h_y(x,y,z) = -\frac{1}{z}, \quad (x,y,z) \in \mathbb{R}^3.$$

Therefore,

$$d\phi(x,y,z) = (g_x(x,y,z) + h_y(x,y,z))dxdydz$$
$$= \left(\frac{1}{z} - \frac{1}{z}\right)dxdydz$$
$$= 0.$$

Thus, ϕ is a closed differential form.

Exercise 5.7. Prove that

$$\phi(x,y,z) = (2xyz^3 + y^2 + 4z + 2)dx$$
$$+ (x^2z^3 + 2xy + 2z^3 - 1)dy$$
$$+ (3x^2yz^2 + 6yz^2 + 4x - 4z)dz$$

is a closed differential form on $\mathbb{R}^3$.

Definition 5.11. A differential form ϕ is said to be exact if there is a closed differential form ψ such that

$$\phi = d\psi.$$

Example 5.19. Let ϕ be as in Example 5.17. Let also

$$\psi(x,y,z) = x^5y^2z^3, \quad (x,y,z) \in \mathbb{R}^3.$$

Then

$$\psi_x(x,y,z) = 5x^4y^2z^3,$$
$$\psi_y(x,y,z) = 2x^5yz^3,$$
$$\psi_z(x,y,z) = 3x^5y^2z^2, \quad (x,y,z) \in \mathbb{R}^3.$$

Thus,

$$\phi(x,y,z) = d\psi(x,y,z), \quad (x,y,z) \in \mathbb{R}^3.$$

Consequently, ϕ is an exact differential form.

Example 5.20. Let ϕ be as in Example 5.18. Let also

$$\psi(x,y,z) = \frac{x^2z^2}{2}dx + z^2dy + \frac{xy}{z}dz, \quad (x,y,z) \in \mathbb{R}^3.$$

Set

$$f(x,y,z) = \frac{x^2z^2}{2},$$
$$g(x,y,z) = z^2,$$
$$h(x,y,z) = \frac{xy}{z}, \quad (x,y,z) \in \mathbb{R}^3.$$

We have

$$g_x(x,y,z) = 0,$$

$$g_z(x, y, z) = 2z,$$
$$f_y(x, y, z) = 0,$$
$$f_z(x, y, z) = x^2 z,$$
$$h_x(x, y, z) = \frac{y}{z},$$
$$h_y(x, y, z) = \frac{x}{z}, \quad (x, y, z) \in \mathbb{R}^3.$$

Therefore,

$$\phi = d\psi.$$

Thus, ϕ is an exact differential form.

Exercise 5.8. Prove that any exact differential form is closed.

Solution. Let ϕ be an exact differential form. Then there is a differential form ψ such that

$$\phi = d\psi.$$

Hence,

$$d\phi = d(d\psi) = 0.$$

Thus, ϕ is a closed differential form. This completes the proof.

Exercise 5.9. Let ϕ be any differential form and ψ be a closed differential form. Prove that

$$d(\phi + \psi) = d\phi.$$

Solution. Since ψ is a closed differential form, we have

$$d\psi = 0.$$

Hence,

$$d(\phi + \psi) = d\phi + d\psi = d\phi.$$

This completes the proof.

Exercise 5.10. Let ϕ_1 and ϕ_2 be two differential forms such that

$$d\phi_1 = d\phi_2.$$

Prove that

$$\phi_2 = \phi_1 + \psi,$$

where ψ is a closed differential form.

Solution. Let

$$\psi = \phi_2 - \phi_1.$$

Then

$$d\psi = d(\phi_2 - \phi_1)$$
$$= d\phi_2 - d\phi_1$$
$$= 0.$$

Thus, ψ is a closed differential form. This completes the proof.

5.5 Gradient, curl, and divergence

Suppose that

$$e_1 = (1, 0, 0),$$
$$e_2 = (0, 1, 0),$$
$$e_3 = (0, 0, 1).$$

Let also Ω be a region in $\mathbb{R}^3$.

Definition 5.12. A vector field on Ω is a vector-valued function

$$F(x, y, z) = f(x, y, z)e_1 + g(x, y, z)e_2 + h(x, y, z)e_3, \quad (x, y, z) \in \Omega,$$

where $f, g, h : \mathbb{R}^3 \to \mathbb{R}$ are given functions.

Example 5.21. Let

$$f(x, y, z) = x^2 - \sin x,$$
$$g(x, y, z) = y + e^z,$$
$$h(x, y, z) = z - \tan(xy), \quad (x, y, z) \in \mathbb{R}^3.$$

Then

$$F(x, y, z) = (x^2 - \sin x)e_1 + (y + e^z)e_2 + (z - \tan(xy))e_3, \quad (x, y, z) \in \mathbb{R}^3,$$

is a vector field.

Definition 5.13. Let $f \in C^1(\Omega)$. Then the gradient of f is defined by

$$\text{grad} f(x,y,z) = f_x(x,y,z)e_1 + f_y(x,y,z)e_2 + f_z(x,y,z)e_3, \quad (x,y,z) \in \Omega.$$

Example 5.22. Let $\Omega = \mathbb{R}^3$ and let

$$f(x,y,z) = x^2 + y^2 + z^2, \quad (x,y,z) \in \mathbb{R}^3.$$

Then

$$f_x(x,y,z) = 2x,$$
$$f_y(x,y,z) = 2y,$$
$$f_z(x,y,z) = 2z, \quad (x,y,z) \in \mathbb{R}^3.$$

Then

$$\text{grad} f(x,y,z) = 2xe_1 + 2ye_2 + 2ze_3, \quad (x,y,z) \in \mathbb{R}^3.$$

Example 5.23. Let $\Omega = \mathbb{R}^3 - \{(x,y,0) : x,y \in \mathbb{R}\}$ and let

$$f(x,y,z) = \frac{x^2 + y^2}{z}, \quad (x,y,z) \in \Omega.$$

Then

$$f_x(x,y,z) = \frac{2x}{z},$$
$$f_y(x,y,z) = \frac{2y}{z},$$
$$f_z(x,y,z) = -\frac{x^2 + y^2}{z^2}, \quad (x,y,z) \in \Omega.$$

Therefore,

$$\text{grad} f(x,y,z) = \frac{2x}{z}e_1 + \frac{2y}{z}e_2 - \frac{x^2 + y^2}{z^2}e_3, \quad (x,y,z) \in \Omega.$$

Exercise 5.11. Let $\Omega = \mathbb{R}^3 - \{(x,y,0) : x,y \in \mathbb{R}\}$ and let

$$f(x,y,z) = \frac{x^3 - 2y^2 + xy}{z^2}, \quad (x,y,z) \in \Omega.$$

Find

$$\text{grad} f(x,y,z), \quad (x,y,z) \in \Omega.$$

Answer 5.7.
$$\text{grad} f(x,y,z) = \frac{3x^2 + y}{z^2} e_1 + \frac{-4y + x}{z^2} e_2 - \frac{2(x^3 - 2y^2 + xy)}{z^3} e_3, \quad (x,y,z) \in \Omega.$$

Definition 5.14. Let

$$F(x,y,z) = f(x,y,z)e_1 + g(x,y,z)e_2 + h(x,y,z)e_3, \quad (x,y,z) \in \Omega, \tag{5.1}$$

where $f, g, h \in C^1(\Omega)$. We define the curl of F as follows:

$$\text{curl} F = (h_y - g_z)e_1 + (f_z - h_x)e_2 + (g_x - f_y)e_3.$$

Example 5.24. Let $\Omega = \mathbb{R}^3$ and let

$$F(x,y,z) = (x^2 + y^2 z)e_1 + (y^2 - 3xz)e_2 + (x^4 + y^3 - z^2)e_3, \quad (x,y,z) \in \mathbb{R}^3.$$

Here

$$f(x,y,z) = x^2 + y^2 z,$$
$$g(x,y,z) = y^2 - 3xz,$$
$$h(x,y,z) = x^4 + y^3 - z^2, \quad (x,y,z) \in \mathbb{R}^3.$$

Then

$$f_y(x,y,z) = 2yz,$$
$$f_z(x,y,z) = y^2,$$
$$g_x(x,y,z) = -3z,$$
$$g_z(x,y,z) = -3x,$$
$$h_x(x,y,z) = 4x^3,$$
$$h_y(x,y,z) = 3y^2, \quad (x,y,z) \in \mathbb{R}^3.$$

Consequently,

$$\text{curl} F(x,y,z) = (3y^2 + 3x)e_1 + (y^2 - 4x^3)e_2 + (-3x - 2yz)e_3, \quad (x,y,z) \in \mathbb{R}^3.$$

Definition 5.15. Let F be given by (5.1). Then its divergence is defined by

$$\text{div} F(x,y,z) = f_x(x,y,z) + g_y(x,y,z) + h_z(x,y,z), \quad (x,y,z) \in \Omega.$$

Example 5.25. Let $\Omega = \mathbb{R}^3$ and

$$F(x,y,z) = x^2 e_1 + y^2 e_2 - z^2 e_3, \quad (x,y,z) \in \mathbb{R}^3.$$

Here

$$f(x,y,z) = x^2,$$
$$g(x,y,z) = y^2,$$
$$h(x,y,z) = -z^2, \quad (x,y,z) \in \mathbb{R}^3.$$

Hence,

$$f_x(x,y,z) = 2x,$$
$$g_y(x,y,z) = 2y,$$
$$h_z(x,y,z) = -2z, \quad (x,y,z) \in \mathbb{R}^3,$$

and

$$\operatorname{div} F(x,y,z) = 2x + 2y - 2z, \quad (x,y,z) \in \mathbb{R}^3.$$

Exercise 5.12. Let $\Omega = \mathbb{R}^3$ and

$$F(x,y,z) = (x^2 + y^2 z)e_1 + xyz e_2 + (x^2 - y^3)e_3, \quad (x,y,z) \in \mathbb{R}^3.$$

Find
1. $\operatorname{curl} F(x,y,z), (x,y,z) \in \mathbb{R}^3$;
2. $\operatorname{div} F(x,y,z), (x,y,z) \in \mathbb{R}^3$.

Answer 5.8. 1.
$$(-3y^2 - xy)e_1 + (y^2 - 2x)e_2 + 2(x - yz)e_3, \quad (x,y,z) \in \mathbb{R}^3;$$

2.
$$2x + xz, \quad (x,y,z) \in \mathbb{R}^3.$$

5.6 Differential forms in $\mathbb{R}^n$

Definition 5.16. A 0-differential form ϕ is a function $f(x_1,\ldots,x_n), (x_1,\ldots,x_n) \in \mathbb{R}^n$.

Definition 5.17. A k-differential form is a sum of terms of the form

$$f(x_1,\ldots,x_n)dx_{j_1} \cdots dx_{j_k}.$$

Addition of differential forms is defined in the usual way. The multiplication of differential forms is subject to the following rules:

$$dx_j dx_k = -dx_k dx_j, \quad j,k \in \{1,\ldots,n\},$$
$$dx_j dx_j = 0, \quad j \in \{1,\ldots,n\}.$$

Definition 5.18. For a 0-differential form ϕ, we define

$$d\phi = \phi_{x_1} dx_1 + \cdots + \phi_{x_n} dx_n.$$

Definition 5.19. Let ϕ be a k-differential form. Its exterior derivative $d\phi$ is the $(k + 1)$-differential form obtained from ϕ by applying d to each function involved in ϕ.

Let

$$e_1 = (1, 0, \ldots, 0),$$
$$e_2 = (0, 1, \ldots, 0),$$
$$\vdots$$
$$e_n = (0, 0, \ldots, 1).$$

Definition 5.20. A vector field on $\mathbb{R}^n$ is a vector-valued function of the form

$$f_1 e_1 + f_2 e_2 + \cdots + f_n e_n,$$

where $f_j : \mathbb{R}^n \to \mathbb{R}, j \in \{1, \ldots, n\}$, are given functions.

5.7 Advanced practical problems

Problem 5.1. Let

$$\phi_1 = x^3 dx + yz dy - (x^2 + y^2 + z^2) dz,$$
$$\phi_2 = y^2 z dx - xz dy + (2x + 1) dz.$$

Find
1. $3\phi_1 - 4\phi_2$;
2. $x\phi_1 + y\phi_2$.

Answer 5.9. 1.

$$(3x^3 - 4y^2 z) dx + (3yz + 4xz) dy - (3x^2 + 3y^2 + 3z^2 + 8x + 4) dz;$$

2.

$$(x^4 + y^3 z) dx + (-x^3 - xy^2 - xz^2 + 2xy + y) dz.$$

Problem 5.2. Let

$$\phi_1 = 2x^2 dx + (x + y) dy,$$
$$\phi_2 = -x dx + (x - 2y) dy,$$
$$\phi_3 = x^3 dx + yz dy - (x^2 + y^2 + z^2) dz,$$

$$\phi_4 = y^2zdx - xzdy + (2x + 1)dz,$$
$$\phi_5 = xdx + y^2dy + z^3dz,$$
$$\phi_6 = dx + 2dy + 3dz.$$

Find

1. $\phi_3\phi_5$;
2. $\phi_3\phi_6$;
3. $\phi_4\phi_5$;
4. $\phi_4\phi_6$;
5. $\phi_5\phi_6$.

Answer 5.10. 1.

$$(x^3y^2 - xyz)dxdy + (yz^4 - y^2(x^2 + y^2 + z^2))dydz - (x^3 + xy^2 + xz^2 - x^3z^3)dxdz;$$

2.

$$(2x^3 - yz)dxdy + (3yz + 2(x^2 + y^2 + z^2))dydz - (x^2 + y^2 + z^2 + 3x^3)dzdx;$$

3.

$$(y^4z + x^2z)dxdy - (xz^4 + 2xy^2 + y^2)dydz + (2x^2 + x - y^2z^4)dzdx;$$

4.

$$(2y^2z + xz)dxdy - (3xz + 4x + 2)dydz + (2x + 1 - 3y^2z)dzdx;$$

5.

$$(2x - y^2)dxdy + (3y^2 - 2z^3)dydz + (z^3 - 3x)dzdx.$$

Problem 5.3. Let $\phi_j, j \in \{1, 2, 3, 4, 5, 6\}$, be as in Problem 5.2. Let also

$$\psi_1 = (x^2 - y^2)dxdy,$$
$$\psi_2 = (x - y)dxdy,$$
$$\psi_3 = (x^2 + y^2)dydz + (x - y^2)dzdx + 3xdxdy,$$
$$\psi_4 = (x^2 - y^2)dydz + (x + y - z^2)dzdx - 6xydxdy.$$

Find

1. $y\psi_1 + x^2\psi_2$;
2. $-\psi_1 + (x + y)\psi_2$;
3. $x\psi_1 + y\psi_2$;
4. $2y\psi_1 + \psi_2$;
5. $\phi_1\psi_1$;
6. $\phi_1\psi_2$;
7. $\phi_2\psi_1$;
8. $\phi_2\psi_2$;
9. $\phi_3\psi_3$;
10. $\phi_3\psi_4$;

11. $\phi_4\psi_3$;
12. $\phi_4\psi_4$;
13. $\phi_1\psi_3$;
14. $\phi_1\psi_4$;
15. $\phi_2\psi_3$;
16. $\phi_2\psi_4$;
17. $\phi_3\psi_1$;
18. $\phi_3\psi_2$;
19. $\phi_4\psi_1$;
20. $\phi_4\psi_2$.

Answer 5.11. 1. $(x^3 - y^3)dxdy$;
2. 0;
3. $(x - y)(x^2 + xy + y)dxdy$;
4. $(x - y)(2y^2 + 2xy + 1)dxdy$;
5. 0;
6. 0;
7. 0;
8. 0;
9. $(x^3(x^2 + y^2) + yz(x - y^2) - 3x(x^2 + y^2 + z^2))dxdydz$;
10. $(x^3(x^2 - y^2) + yz(x + y - z^2) + 6xy(x^2 + y^2 + z^2))dxdydz$;
11. $(y^2z(x^2 + y^2) - xz(x - y^2) + 3x(2x + 1))dxdydz$;
12. $(y^2z(x^2 - y^2) - xz(x + y - z^2) - 6xy(2x + 1))dxdydz$;
13. $(2x^2(x^2 + y^2) + (x + y)(x^2 - y^2))dxdydz$;
14. $(2x^2(x^2 - y^2) + (x + y)(x + y - z^2))dxdydz$;
15. $(-x(x^2 + y^2) + (x - 2y)(x - y^2))dxdydz$;
16. $(-x(x^2 - y^2) + (x - 2y)(x + y - z^2))dxdydz$;
17. $-(x^2 + y^2 + z^2)(x^2 - y^2)dxdydz$;
18. $-(x^2 + y^2 + z^2)(x - y)dxdydz$;
19. $(2x + 1)(x^2 - y^2)dxdydz$;
20. $(2x + 1)(x - y)dxdydz$.

Problem 5.4. Let $\phi_j, j \in \{1, \ldots, 6\}$, be as in Problem 5.2 and

$$\phi_7 = x^2yzdx - 2xy^3zdy + xyz^4dz,$$
$$\phi_8 = yz^2dx + (x + z^2)dy + (x^2 - y)dz, \quad (x, y, z) \in \mathbb{R}^3.$$

Find
1. $d\phi_1$;
2. $d\phi_2$;
3. $d\phi_3$;
4. $d\phi_4$;

5. $d\phi_5$;
6. $d\phi_6$;
7. $d\phi_7$;
8. $d\phi_8$.

Answer 5.12. 1. $dxdy$;
2. $dxdy$;
3. $-3ydydz + 2xdzdx$;
4. $-(z + 2yz)dxdy + xdydz + (y^2 - 2)dzdx$;
5. 0;
6. 0;
7. $(2x^3 + xz^4)dydz + (x^2y - yz^4)dzdx - (x^2z + 2y^3z)dxdy$;
8. $(1 - z^2)dxdy - (2z + 1)dydz + (2yz - 2x)dzdx$.

Problem 5.5. Let $\psi_j, j \in \{1, \ldots, 4\}$, be as in Problem 5.3 and

$$\psi_5 = x^2yzdydz + 2xyzdzdx + xyz^3dxdy,$$
$$\psi_6 = (x + 2y - 3z)dydz - (3x + 4yz^2)dzdx + (9xz^2 - 4xy)dxdy.$$

Find
1. $d\psi_1$;
2. $d\psi_2$;
3. $d\psi_3$;
4. $d\psi_4$;
5. $d\psi_5$;
6. $d\psi_6$.

Answer 5.13. 1. $0, \quad (x, y, z) \in \mathbb{R}^3$;
2. $0, \quad (x, y, z) \in \mathbb{R}^3$;
3. $(2x - 2y)dxdydz, \quad (x, y, z) \in \mathbb{R}^3$;
4. $(2x + 1)dxdydz, \quad (x, y, z) \in \mathbb{R}^3$;
5. $(2xyz - 2xz + 3xyz^2)dxdydz, \quad (x, y, z) \in \mathbb{R}^3$;
6. $(1 - 4z^2 + 18xz)dxdydz, \quad (x, y, z) \in \mathbb{R}^3$.

Problem 5.6. Let $\phi_j, j \in \{1, \ldots, 4\}$, be as in Problem 5.2. Find
1. $d(\phi_1\phi_2)$;
2. $d(\phi_1\phi_3)$;
3. $d(\phi_1\phi_4)$;
4. $d(\phi_2\phi_3)$;
5. $d(\phi_2\phi_4)$;
6. $d(\phi_3\phi_4)$.

Answer 5.14. 1. $0, \quad (x, y, z) \in \mathbb{R}^3$;

2. $(-(x^2 + y^2 + z^2) - 2x(x + y) + 6x^2y)dxdydz, \quad (x, y, z) \in \mathbb{R}^3;$
3. $(4x + 2y + 1 - 2x^3 - xy^2 - y^3)dxdydz, \quad (x, y, z) \in \mathbb{R}^3;$
4. $-(3x^2 + y^2 + z^2 - xy)dxdydz, \quad (x, y, z) \in \mathbb{R}^3;$
5. $(4x - 4y + 1 + x^2 - xy^2 + 2y^3)dxdydz, \quad (x, y, z) \in \mathbb{R}^3;$
6. $(2yz - 3x^2z - y^2z - z^3 - 2x^2yz - 6y^3z - 2yz^3 - x^4)dxdydz, \quad (x, y, z) \in \mathbb{R}^3.$

Problem 5.7. Let $\phi_j, j \in \{1, \ldots, 6\}$, be as in Problem 5.2. Check
1. $d(\phi_j\phi_l) = (d\phi_j)\phi_l + \phi_j d\phi_l, j, l \in \{1, \ldots, 6\};$
2. $d^2\phi_j = 0, j \in \{1, \ldots, 6\}.$

Problem 5.8. Prove that the differential form

$$\phi = -\frac{y}{x^2 + y^2}dx + \frac{x}{x^2 + y^2}dy, \quad (x, y, z) \in \mathbb{R}^3 - \{(0, 0, z) : z \in \mathbb{R}\},$$

is closed, but not exact.

Problem 5.9. Prove that

$$\phi = \frac{x}{(x^2 + y^2 + z^2)^{\frac{3}{2}}}dydz + \frac{y}{(x^2 + y^2 + z^2)^{\frac{3}{2}}}dzdx + \frac{z}{(x^2 + y^2 + z^2)^{\frac{3}{2}}}dxdy,$$

for $(x, y, z) \in \mathbb{R}^3 \setminus \{(0, 0, 0)\}$, is closed, but not exact.

Problem 5.10. Let

$$f_1(x, y, z) = x^2y^3z - 2xyz^2,$$
$$f_2(x, y, z) = x^2 + y^2 - 3z^4,$$
$$f_3(x, y, z) = x^2y + y^2z - z^2x,$$
$$f_4(x, y, z) = 2x + 3y - 4z + 5, \quad (x, y, z) \in \mathbb{R}^3.$$

Find
1. $\mathrm{grad}f_1;$
2. $\mathrm{grad}f_2;$
3. $\mathrm{grad}f_3;$
4. $\mathrm{grad}f_4.$

Answer 5.15. 1.

$$(2xy^3z - 2yz^2)e_1 + (3x^2y^2z - 2xz^2)e_2 + (x^2y^3 - 4xyz)e_3, \quad (x, y, z) \in \mathbb{R}^3.$$

2.
$$2xe_1 + 2ye_2 - 12z^3e_3, \quad (x, y, z) \in \mathbb{R}^3;$$

3.
$$(2xy - z^2)e_1 + (x^2 + 2yz)e_2 + (y^2 - 2xz)e_3, \quad (x, y, z) \in \mathbb{R}^3;$$

4.
$$2e_1 + 3e_2 - 4e_3, \quad (x, y, z) \in \mathbb{R}^3.$$

Problem 5.11. Let

$$F_1(x, y, z) = 2x^2e_1 + (x + y)e_2,$$
$$F_2(x, y, z) = -xe_1 + (x - 2y)e_2,$$
$$F_3(x, y, z) = x^2e_1 + yze_2 - (x^2 + y^2 + z^2)e_3,$$
$$F_4(x, y, z) = y^2ze_1 - xze_2 + (2x + 1)e_3,$$
$$F_5(x, y, z) = xe_1 + y^2e_2 + z^3e_3,$$
$$F_6(x, y, z) = e_1 + 2e_2 + 3e_3,$$
$$F_7(x, y, z) = x^2yze_1 - 2xy^3ze_2 + 3xyz^4e_3,$$
$$F_8(x, y, z) = y^2ze_1 + (x + z^2)e_2 + (x^2 - y)e_3, \quad (x, y, z) \in \mathbb{R}^3.$$

Find
1. $\operatorname{curl} F_1$;
2. $\operatorname{curl} F_2$;
3. $\operatorname{curl} F_3$;
4. $\operatorname{curl} F_4$;
5. $\operatorname{curl} F_5$;
6. $\operatorname{curl} F_6$;
7. $\operatorname{curl} F_7$;
8. $\operatorname{curl} F_8$.

Answer 5.16. 1. $e_3, \quad (x, y, z) \in \mathbb{R}^3$;
2. $e_3, \quad (x, y, z) \in \mathbb{R}^3$;
3. $-3ye_1 + 2xe_2, \quad (x, y, z) \in \mathbb{R}^3$;
4. $xe_1 + (y^2 - 2)e_2 + (z - 2yz)e_3, \quad (x, y, z) \in \mathbb{R}^3$;
5. $0, \quad (x, y, z) \in \mathbb{R}^3$;
6. $0, \quad (x, y, z) \in \mathbb{R}^3$;
7. $(2xy^3 + 3xz^4)e_1 + (x^2y - 3yz^4)e_2 - (x^2z + 2y^3z)e_3, \quad (x, y, z) \in \mathbb{R}^3$;
8. $(1 - 2yz)e_1 + (y^2 - 2x)e_2 + (1 - z^2)e_3, \quad (x, y, z) \in \mathbb{R}^3$.

Problem 5.12. Let

$$G_1(x, y, z) = (x^2 - y^2)e_3,$$
$$G_2(x, y, z) = (x - y)e_3,$$
$$G_3(x, y, z) = (x^2 + y^2)e_1 + (x - y^2)e_2 + 3xe_3,$$
$$G_4(x, y, z) = (x - y^2)e_1 + (x + y - z^2)e_2 - 6xye_3,$$
$$G_5(x, y, z) = x^2yze_1 - 2xyze_2 + xyz^3e_3,$$

$$G_6(x, y, z) = (x + 2y - 3z)e_1 - (3x + 4yz^2)e_2 + (9xz^2 - 4xy)e_3, \quad (x, y, z) \in \mathbb{R}^3.$$

Find
1. $\operatorname{div} G_1$;
2. $\operatorname{div} G_2$;
3. $\operatorname{div} G_3$;
4. $\operatorname{div} G_4$;
5. $\operatorname{div} G_5$;
6. $\operatorname{div} G_6$.

Answer 5.17. 1. $0, \quad (x, y, z) \in \mathbb{R}^3.$
2. $0, \quad (x, y, z) \in \mathbb{R}^3$;
3. $2x - 2y, \quad (x, y, z) \in \mathbb{R}^3$;
4. $2, \quad (x, y, z) \in \mathbb{R}^3$;
5. $2xyz - 2xz + 3xyz^2, \quad (x, y, z) \in \mathbb{R}^3$;
6. $1 - 4z^2 + 18xz, \quad (x, y, z) \in \mathbb{R}^3$.

6 The nature connection

In this chapter we first define the notions of directional and covariant derivatives acting on the set of vector fields. Analogously, the Lie brackets are also considered.

6.1 Directional derivatives

Let $x^0 = (x_1^0, \ldots, x_n^0) \in \mathbb{R}^n$ and v be a nonzero vector in $\mathbb{R}^n$ such that

$$v = v_1 e_1 + \cdots + v_n e_n.$$

Let also $\theta_j = \angle(v, e_j), j = 1, \ldots, n.$ Then

$$\cos \theta_j = \frac{v_j}{|v|} = \frac{v_j}{\sqrt{v_1^2 + \cdots + v_n^2}}, \quad j = 1, \ldots, n.$$

We have that

$$(\cos \theta_1)^2 + \cdots + (\cos \theta_n)^2 = 1.$$

Set $v_0 = \frac{v}{|v|}$. Then

$$v_0 = \cos \theta_1 e_1 + \cdots + \cos \theta_n e_n.$$

Let l be a line through the point x^0 and parallel to v. Then

$$l : \quad \begin{aligned} x_1 &= x_1^0 + t \cos \theta_1, \\ &\vdots \\ x_n &= x_n^0 + t \cos \theta_n, \end{aligned}$$

where $t > 0$ is the distance between the points $x = (x_1, \ldots, x_n) \in l$ and x^0. We have indeed

$$\begin{aligned} d(x, x^0) &= \sqrt{(t \cos \theta_1)^2 + \cdots + (t \cos \theta_n)^2} \\ &= \sqrt{t^2((\cos \theta_1)^2 + \cdots + (\cos \theta_n)^2)} \\ &= \sqrt{t^2} \\ &= t. \end{aligned}$$

Suppose that the function f is defined in a neighborhood of the point x^0.

https://doi.org/10.1515/9783111501857-006

Definition 6.1. The derivative of f at x^0 in the direction v is defined by

$$\left.\frac{\partial f}{\partial v}\right|_{x^0} = \lim_{x \to x^0} \frac{f(x) - f(x^0)}{d(x, x^0)}.$$

Let f be differentiable at x^0. Then

$$\left.\frac{\partial f}{\partial v}\right|_{x^0} = \lim_{t \to 0} \frac{f(x_1^0 + t \cos \theta_1, \ldots, x_n^0 + t \cos \theta_n) - f(x_1^0, \ldots, x_n^0)}{t}$$

$$= \left.\frac{d}{dt} f(x_1^0 + t \cos \theta_1, \ldots, x_n^0 + t \cos \theta_n)\right|_{t=0}$$

$$= f_{x_1}(x^0) \cos \theta_1 + \cdots + f_{x_n}(x^0) \cos \theta_n$$

$$= \langle \mathrm{grad} f(x^0), v_0 \rangle.$$

Example 6.1. We will find $\left.\frac{\partial f}{\partial v}\right|_p$, where

$$f(x_1, x_2) = 3x_1^2 + 5x_2^2, \quad (x_1, x_2) \in \mathbb{R}^2,$$

$$v = \left(-\frac{1}{\sqrt{2}}, \frac{1}{\sqrt{2}}\right), \quad p(1, 1).$$

We have

$$|v| = \sqrt{\left(-\frac{1}{\sqrt{2}}\right)^2 + \left(\frac{1}{\sqrt{2}}\right)^2}$$

$$= \sqrt{\frac{1}{2} + \frac{1}{2}}$$

$$= 1.$$

Then,

$$\cos \theta_1 = -\frac{1}{\sqrt{2}},$$

$$\cos \theta_2 = \frac{1}{\sqrt{2}}.$$

Next,

$$f_{x_1}(x_1, x_2) = 6x_1,$$

$$f_{x_2}(x_1, x_2) = 10x_2, \quad (x_1, x_2) \in \mathbb{R}^2.$$

Hence,

$$f_{x_1}(p) = f_{x_1}(1, 1) = 6,$$

$$f_{x_2}(p) = f_{x_2}(1, 1) = 10,$$

and

$$\frac{\partial f}{\partial v}\Big|_p = f_{x_1}(p) \cos \theta_1 + f_{x_2}(p) \cos \theta_2$$

$$= 6\left(-\frac{1}{\sqrt{2}}\right) + 10\left(\frac{1}{\sqrt{2}}\right)$$

$$= 6\left(-\frac{\sqrt{2}}{2}\right) + 10\left(\frac{\sqrt{2}}{2}\right)$$

$$= 2\sqrt{2}.$$

Example 6.2. We will find $\frac{\partial f}{\partial v}\big|_p$, where

$$f(x_1, x_2, x_3) = x_1^3 + 2x_1x_2^2 + 3x_2x_3^2, \quad (x_1, x_2, x_3) \in \mathbb{R}^3,$$

$$v = \left(\frac{2}{3}, \frac{2}{3}, \frac{1}{3}\right), \quad p(3, 3, 1).$$

We have

$$|v| = \sqrt{\left(\frac{2}{3}\right)^2 + \left(\frac{2}{3}\right)^2 + \left(\frac{1}{3}\right)^2}$$

$$= \sqrt{\frac{4}{9} + \frac{4}{9} + \frac{1}{9}}$$

$$= 1.$$

Thus,

$$\cos \theta_1 = \frac{2}{3},$$

$$\cos \theta_2 = \frac{2}{3},$$

$$\cos \theta_3 = \frac{1}{3}.$$

Next,

$$f_{x_1}(x_1, x_2, x_3) = 3x_1^2 + 2x_2^2,$$

$$f_{x_2}(x_1, x_2, x_3) = 4x_1x_2 + 3x_3^2,$$

$$f_{x_3}(x_1, x_2, x_3) = 6x_2x_3, \quad (x_1, x_2, x_3) \in \mathbb{R}^3.$$

Hence,

$$f_{x_1}(p) = f_{x_1}(3, 3, 1)$$

$$= 3 \cdot 3^2 + 2 \cdot 3^2$$

$$= 45,$$

$$f_{x_2}(p) = f_{x_2}(3,3,1)$$
$$= 4 \cdot 3 \cdot 3$$
$$= 39,$$
$$f_{x_3}(p) = f_{x_3}(3,3,1)$$
$$= 6 \cdot 3 \cdot 1$$
$$= 18,$$

and

$$\left.\frac{\partial f}{\partial v}\right|_p = f_{x_1}(p) \cos \theta_1 + f_{x_2}(p) \cos \theta_2 + f_{x_3}(p) \cos \theta_3$$
$$= 45 \cdot \frac{2}{3} + 39 \cdot \frac{2}{3} + 18 \cdot \frac{1}{3}$$
$$= 30 + 26 + 6$$
$$= 62.$$

Example 6.3. We will find $\left.\frac{\partial f}{\partial v}\right|_p$, where

$$f(x_1,\ldots,x_n) = \sum_{j=1}^n \arcsin x_j, \quad (x_1,\ldots,x_n) \in \mathbb{R}^n, \quad |(x_1,\ldots,x_n)| < 1,$$

$$v = \left(\frac{1}{\sqrt{n}},\ldots,\frac{1}{\sqrt{n}}\right), \quad p\left(\frac{1}{4},\ldots,\frac{1}{4}\right).$$

We have

$$|v| = \sqrt{\left(\frac{1}{\sqrt{n}}\right)^2 + \cdots + \left(\frac{1}{\sqrt{n}}\right)^2}$$
$$= \sqrt{\frac{1}{n} + \cdots + \frac{1}{n}}$$
$$= 1.$$

Thus,

$$\cos \theta_j = \frac{1}{\sqrt{n}}, \quad j = 1,\ldots,n.$$

Next,

$$f_{x_j}(x_1,\ldots,x_n) = \frac{1}{\sqrt{1-x_j^2}}, \quad (x_1,\ldots,x_n) \in \mathbb{R}^n, \quad |(x_1,\ldots,x_n)| < 1.$$

Therefore,

$$f_{x_j}(p) = f_{x_j}\left(\frac{1}{4}, \ldots, \frac{1}{4}\right)$$

$$= \frac{1}{\sqrt{1 - \frac{1}{16}}}$$

$$= \frac{4}{\sqrt{15}}, \quad j = 1, \ldots, n,$$

and

$$\left.\frac{\partial f}{\partial v}\right|_p = \sum_{j=1}^{n} f_{x_j}(p) \cos \theta_j$$

$$= \sum_{j=1}^{n} \frac{4}{\sqrt{15}} \cdot \frac{1}{\sqrt{n}}$$

$$= \frac{4n}{\sqrt{15n}}$$

$$= 4\sqrt{\frac{n}{15}}.$$

Exercise 6.1. Find $\left.\frac{\partial f}{\partial v}\right|_p$, where

1.

$$f(x_1, x_2) = x_1 \sin(x_1 + x_2), \quad (x_1, x_2) \in \mathbb{R}^2,$$

$$v = (-1, 0), \quad p\left(\frac{\pi}{4}, \frac{\pi}{4}\right).$$

2.

$$f(x_1, x_2, x_3) = \log(x_1^2 + x_2^2 + x_3^2), \quad (x_1, x_2, x_3) \in \mathbb{R}^3, \quad x_1^2 + x_2^2 + x_3^2 \neq 0,$$

$$v = \left(-\frac{1}{3}, \frac{2}{3}, \frac{2}{3}\right), \quad p(1, 2, 3).$$

3.

$$f(x_1, x_2, x_3, x_4) = x_1^2 + x_2^2 - x_3^2 + x_4^2, \quad (x_1, x_2, x_3, x_4) \in \mathbb{R}^4,$$

$$v = \left(\frac{2}{3}, \frac{1}{3}, 0, -\frac{2}{3}\right), \quad p(1, 3, 2, 1).$$

Answer 6.1. 1. $\quad -1$;

2. $\frac{3}{7}$;

3. $\quad 2$.

Exercise 6.2. Let f and g be differentiable at x^0. Prove that

1.

$$\left.\frac{\partial(af + bg)}{\partial v}\right|_{x^0} = a\left.\frac{\partial f}{\partial v}\right|_{x^0} + b\left.\frac{\partial g}{\partial v}\right|_{x^0},$$

for any $a, b \in \mathbb{R}$;

2.

$$\frac{\partial (fg)}{\partial v}\bigg|_{x^0} = g(x_0)\frac{\partial f}{\partial v}\bigg|_{x^0} + f(x_0)\frac{\partial g}{\partial v}\bigg|_{x^0}.$$

Exercise 6.3 (Euler identity). Let $\Omega \subset \mathbb{R}^n, f : \Omega \to \mathbb{R}$ be a homogeneous function with degree of homogeneity α, and $f \in C^1(\Omega)$. Then

$$\alpha f(x_1,\ldots,x_n) = x_1 f_{x_1}(x_1,\ldots,x_n) + \cdots + x_n f_{x_n}(x_1,\ldots,x_n), \quad (x_1,\ldots,x_n) \in \Omega. \tag{6.1}$$

Definition 6.2. The identity (6.1) is said to be the Euler identity.

Solution. Since f is a homogeneous function with degree of homogeneity α on Ω, for any $t \in \mathbb{R}$ and $(x_1,\ldots,x_n) \in \Omega$ we have $(tx_1,\ldots,tx_n) \in \Omega$ such that

$$f(tx_1,\ldots,tx_n) = t^\alpha f(x_1,\ldots,x_n), \quad (x_1,\ldots,x_n) \in \Omega.$$

We differentiate with respect to t and find

$$\alpha t^{\alpha-1} f(x_1,\ldots,x_n) = x_1 \frac{\partial f}{\partial (tx_1)}(tx_1,\ldots,tx_n) + \cdots + x_n \frac{\partial f}{\partial (tx_n)}(tx_1,\ldots,tx_n).$$

Putting $t = 1$ in the latter equation, we find (6.1). This completes the proof.

6.2 Tangent spaces

Let $T_p(\mathbb{R}^n)$ be the tangent space with a point of application $p \in \mathbb{R}^n$ and $(x_1,\ldots,x_n)$ be the canonical coordinates of $\mathbb{R}^n$.

Definition 6.3. The set

$$\left\{ \frac{\partial}{\partial x_1}\bigg|_p, \ldots, \frac{\partial}{\partial x_n}\bigg|_p \right\}$$

is said to be the basis of $T_p(\mathbb{R}^n)$.

Any vector field V can be represented as

$$V = \sum_{j=1}^n V_j \frac{\partial}{\partial x_j}.$$

Definition 6.4. Let

$$V = \sum_{j=1}^n V_j \frac{\partial}{\partial x_j},$$

$$W = \sum_{j=1}^n W_j \frac{\partial}{\partial x_j},$$

where $V_j, W_j : \mathbb{R}^n \to \mathbb{R}, j = 1,\ldots,n$. Let also $f : \mathbb{R}^n \to \mathbb{R}$. We define

$$aV + bW = \sum_{j=1}^{n}(aV_j + bW_j)\frac{\partial}{\partial x_j}, \quad a, b \in \mathbb{R},$$

and

$$fV = \sum_{j=1}^{n}(fV_j)\frac{\partial}{\partial x_j}.$$

Example 6.4. Let

$$V = (3 + x_1 + x_2)\frac{\partial}{\partial x_1} + x_1^2\frac{\partial}{\partial x_2},$$

$$W = (x_1 - x_2)\frac{\partial}{\partial x_1} + \frac{\partial}{\partial x_2}.$$

Then

$$V + W = (3 + 2x_1)\frac{\partial}{\partial x_1} + (x_1^2 + 1)\frac{\partial}{\partial x_2}$$

and

$$x_1 V = (3x_1 + x_1^2 + x_1 x_2)\frac{\partial}{\partial x_1} + x_1^3\frac{\partial}{\partial x_2}.$$

Exercise 6.4. Let

$$V = x_1\frac{\partial}{\partial x_1} + x_1\frac{\partial}{\partial x_2},$$

$$W = (x_1 - x_2)\frac{\partial}{\partial x_1} + x_1\frac{\partial}{\partial x_2}.$$

Find
1. $V + W$;
2. $V - W$;
3. $(x_1 + x_2)V$;
4. $x_2 W$.

Answer 6.2. 1.

$$(2x_1 - x_2)\frac{\partial}{\partial x_1} + 2x_1\frac{\partial}{\partial x_2};$$

2.

$$x_2\frac{\partial}{\partial x_1};$$

3.
$$x_1(x_1 + x_2)\left(\frac{\partial}{\partial x_1} + \frac{\partial}{\partial x_2}\right);$$

4.
$$(x_1 x_2 - x_2^2)\frac{\partial}{\partial x_1} + x_1 x_2 \frac{\partial}{\partial x_2}.$$

6.3 Covariant derivatives

Let $p \in \mathbb{R}^n$, $v \in T_p(\mathbb{R}^n)$, and

$$W = \sum_{j=1}^{n} w_j \frac{\partial}{\partial x_j},$$

where w_j is differentiable at p.

Definition 6.5. The covariant derivative of W with respect to v is defined by

$$\nabla_v W = \sum_{j=1}^{n} \left.\frac{\partial w_j}{\partial v}\right|_p \left.\frac{\partial}{\partial x_j}\right|_p.$$

Exercise 6.5. Let

$$W_1 = \sum_{j=1}^{n} w_{1j} \frac{\partial}{\partial x_j},$$

$$W_2 = \sum_{j=1}^{n} w_{2j} \frac{\partial}{\partial x_j},$$

and $a, b \in \mathbb{R}$. Prove that

$$\nabla_v(aW_1 + bW_2) = a\nabla_v W_1 + b\nabla_v W_2.$$

Solution. We have

$$\nabla_v W_1 = \sum_{j=1}^{n} \left.\frac{\partial w_{1j}}{\partial v}\right|_p \left.\frac{\partial}{\partial x_j}\right|_p,$$

$$\nabla_v W_2 = \sum_{j=1}^{n} \left.\frac{\partial w_{2j}}{\partial v}\right|_p \left.\frac{\partial}{\partial x_j}\right|_p,$$

and

$$aW_1 + bW_2 = \sum_{j=1}^{n} (aw_{1j} + bw_{2j})\frac{\partial}{\partial x_j}.$$

Then

$$\nabla_v(aW_1 + bW_2) = \sum_{j=1}^n \frac{\partial(aw_{1j} + bw_{2j})}{\partial v}\bigg|_p \frac{\partial}{\partial x_j}\bigg|_p$$

$$= \sum_{j=1}^n \left(\frac{a\partial w_{1j}}{\partial v}\bigg|_p + b\frac{\partial w_{2j}}{\partial v}\bigg|_p\right)\frac{\partial}{\partial x_j}\bigg|_p$$

$$= \sum_{j=1}^n \left(a\frac{\partial w_{1j}}{\partial v}\bigg|_p \frac{\partial}{\partial x_j}\bigg|_p + b\frac{\partial w_{2j}}{\partial v}\bigg|_p \frac{\partial}{\partial x_j}\bigg|_p\right)$$

$$= \sum_{j=1}^n a\frac{\partial w_{1j}}{\partial v}\bigg|_p \frac{\partial}{\partial x_j}\bigg|_p + \sum_{j=1}^n b\frac{\partial w_{2j}}{\partial v}\bigg|_p \frac{\partial}{\partial x_j}\bigg|_p$$

$$= a\sum_{j=1}^n \frac{\partial w_{1j}}{\partial v}\bigg|_p \frac{\partial}{\partial x_j}\bigg|_p + b\sum_{j=1}^n \frac{\partial w_{2j}}{\partial v} \frac{\partial}{\partial x_j}\bigg|_p$$

$$= a\nabla_v W_1 + b\nabla_v W_2.$$

This completes the proof.

Exercise 6.6. Let

$$W = \sum_{j=1}^n w_j \frac{\partial}{\partial x_j}$$

and $f : \mathbb{R}^n \to \mathbb{R}, f \in C^1(\mathbb{R}^n)$. Prove that

$$\nabla_v(fW) = f\nabla_v W + \frac{\partial f}{\partial v}\bigg|_p W.$$

Solution. We have

$$\nabla_v W = \sum_{j=1}^n \frac{\partial w_j}{\partial v}\bigg|_p \frac{\partial}{\partial x_j}\bigg|_p$$

and

$$fW = \sum_{j=1}^n (fw_j)\frac{\partial}{\partial x_j}.$$

Then

$$\nabla_v(fW) = \sum_{j=1}^n \frac{\partial(fw_j)}{\partial v}\bigg|_p \frac{\partial}{\partial x_j}\bigg|_p$$

$$= \sum_{j=1}^n \left(f\frac{\partial w_j}{\partial v}\bigg|_p + w_j(p)\frac{\partial f}{\partial v}\bigg|_p\right)\frac{\partial}{\partial x_j}\bigg|_p$$

$$= \sum_{j=1}^{n} \left(f \frac{\partial w_j}{\partial v} \Big|_p \frac{\partial}{\partial x_j} + w_j(p) \frac{\partial f}{\partial v} \frac{\partial}{\partial x_j} \Big|_p \right)$$

$$= \sum_{j=1}^{n} f \frac{\partial w_j}{\partial v} \Big|_p \frac{\partial}{\partial x_j} \Big|_p + \sum_{j=1}^{n} \frac{\partial f}{\partial v} \Big|_p w_j(p) \frac{\partial}{\partial x_j} \Big|_p$$

$$= f \sum_{j=1}^{n} \frac{\partial w_j}{\partial v} \Big|_p \frac{\partial}{\partial x_j} \Big|_p + \frac{\partial f}{\partial v} \Big|_p \sum_{j=1}^{n} w_j(p) \frac{\partial}{\partial x_j} \Big|_p$$

$$= f \nabla_v W + \frac{\partial f}{\partial v} \Big|_p W(p).$$

This completes the proof.

Example 6.5. Let $n = 2$ and

$$W(x_1, x_2) = (x_1^2 + x_2) \frac{\partial}{\partial x_1} + 2x_1 x_2 \frac{\partial}{\partial x_2}, \quad (x_1, x_2) \in \mathbb{R}^2.$$

Also, let $p = (1, 1)$ and $v = (-\frac{1}{\sqrt{2}}, \frac{1}{\sqrt{2}}) \in T_p \mathbb{R}^2$. Here

$$w_1(x_1, x_2) = x_1^2 + x_2,$$
$$w_2(x_1, x_2) = 2x_1 x_2, \quad (x_1, x_2) \in \mathbb{R}^2.$$

Then

$$w_{1x_1}(x_1, x_2) = 2x_1,$$
$$w_{1x_2}(x_1, x_2) = 1,$$
$$w_{2x_1}(x_1, x_2) = 2x_2,$$
$$w_{2x_2}(x_1, x_2) = 2x_1, \quad (x_1, x_2) \in \mathbb{R}^2,$$

and

$$\frac{\partial w_1}{\partial v} \Big|_p = (w_{1x_1} v_1 + w_{1x_2} v_2)(p)$$

$$= \left(\left(-\frac{1}{\sqrt{2}} \right) (2x_1) + \frac{1}{\sqrt{2}} \right) (p)$$

$$= \left(-\frac{1}{\sqrt{2}} (2x_1 - 1) \right) (p),$$

$$= -\frac{1}{\sqrt{2}},$$

as well as

$$\left.\frac{\partial w_2}{\partial v}\right|_p = (w_{2x_1} v_1 + w_{2x_2} v_2)(p)$$

$$= \left(\left(-\frac{1}{\sqrt{2}}\right)(2x_2) + \frac{1}{\sqrt{2}}(2x_1)\right)(p)$$

$$= -\sqrt{2}(x_2 - x_1)(p),$$

$$= 0.$$

Hence,

$$\nabla_v W = \left.\frac{\partial w_1}{\partial v}\right|_p \frac{\partial}{\partial x_1} + \left.\frac{\partial w_2}{\partial v}\right|_p \frac{\partial}{\partial x_2}$$

$$= -\frac{1}{\sqrt{2}} \left.\frac{\partial}{\partial x_1}\right|_p,$$

or

$$\nabla_v W = \left(-\frac{1}{\sqrt{2}}, 0\right).$$

Exercise 6.7. Let $n = 2$ and $v = (-\frac{1}{\sqrt{2}}, \frac{1}{\sqrt{2}})$. Find $\nabla_v W$, where

1.

$$W(x_1, x_2) = (x_1^3 + x_2^3 - 3x_1x_2)\frac{\partial}{\partial x_1} + \frac{x_1(x_1 - x_2)}{x_2^2}\frac{\partial}{\partial x_2}, \quad (x_1, x_2) \in \mathbb{R}^2;$$

2.

$$W(x_1, x_2) = (\sin x_1 - x_1^2 x_2)\frac{\partial}{\partial x_1} + \sin\left(\frac{x_1}{x_2}\right)\cos\left(\frac{x_2}{x_1}\right)\frac{\partial}{\partial x_2}, \quad (x_1, x_2) \in \mathbb{R}^2;$$

3.

$$W(x_1, x_2) = e^{x_1}(\cos x_2 + x_1 \sin x_2)\frac{\partial}{\partial x_1} + \log\left(\frac{\sqrt{x_1^2 + x_2^2} - x_1}{\sqrt{x_1^2 + x_2^2} + x_1}\right)\frac{\partial}{\partial x_2}, \quad (x_1, x_2) \in \mathbb{R}^2;$$

4.

$$W(x_1, x_2) = \arcsin\left(\sqrt{\frac{x_1^2 - x_2^2}{x_1^2 + x_2^2}}\right)\frac{\partial}{\partial x_1} + (1 + (\sin x_1)^2)^{\log x_2}\frac{\partial}{\partial x_2}, \quad (x_1, x_2) \in \mathbb{R}^2.$$

Answer 6.3. 1.

$$\nabla_v W(x_1, x_2) = \frac{3}{\sqrt{2}}(x_2^2 + x_2 - x_1 - x_1^2)\frac{\partial}{\partial x_1} + \frac{x_2^2 - x_1x_2 - 2x_1^2}{\sqrt{2}x_2^3}\frac{\partial}{\partial x_2}, \quad (x_1, x_2) \in \mathbb{R}^2;$$

2.

$$\nabla_v W(x_1, x_2) = \frac{1}{\sqrt{2}}e^{x_1}(x_1(\cos x_2 - \sin x_2) - 2\sin x_2 - \cos x_2)\frac{\partial}{\partial x_1}$$

$$+ \sqrt{2}\frac{x_1 + x_2}{x_2\sqrt{x_1^2 + x_2^2}}\frac{\partial}{\partial x_2}, \quad (x_1, x_2) \in \mathbb{R}^2;$$

3.

$$\nabla_l W(x_1, x_2) = \frac{1}{\sqrt{2}} \frac{x_1 x_2 (x_1 |x_1| + x_2 |x_2|) \sqrt{2x_1^2 - 2x_2^2}}{|x_1||x_2|(x_2^4 - x_1^4)} \frac{\partial}{\partial x_1}$$
$$+ \frac{1}{\sqrt{2}} \left(\sin(2x_1) \log x_2 (4 + (\sin x_1)^2)^{\log x_2 - 1} \right.$$
$$\left. - \frac{1}{x_2} (1 + (\sin x_1)^2)^{\log x_2} \log(1 + (\sin x_1)^2) \right) \frac{\partial}{\partial x_2}, \quad (x_1, x_2) \in \mathbb{R}^2.$$

6.4 The Lie brackets

Suppose that

$$V_1 = \sum_{j=1}^{n} V_{1j} \frac{\partial}{\partial x_j},$$

$$V_2 = \sum_{j=1}^{n} V_{2j} \frac{\partial}{\partial x_j},$$

where $V_{kj} : \mathbb{R}^n \to \mathbb{R}$, $V_{kj} \in C^1(\mathbb{R}^n)$, for $k = 1, 2$ and $j = 1, \ldots, n$.

Definition 6.6. Let $f : \mathbb{R}^n \to \mathbb{R}$ be a differentiable function. Define

$$V_1(V_2(f)) = \sum_{j=1}^{n} V_{1j} \frac{\partial}{\partial x_j} \left(\sum_{k=1}^{n} V_{2k} \frac{\partial f}{\partial x_k} \right).$$

Example 6.6. Let

$$V_1(x_1, x_2) = (x_1^2 + x_2^2) \frac{\partial}{\partial x_1} + x_1 x_2 \frac{\partial}{\partial x_2},$$

$$V_2(x_1, x_2) = (x_1^2 - x_2^2) \frac{\partial}{\partial x_1} + (x_1 + x_2) \frac{\partial}{\partial x_2},$$

$$f(x_1, x_2) = (x_1^3 + x_2^3), \quad (x_1, x_2) \in \mathbb{R}^2.$$

Here

$$V_{11}(x_1, x_2) = x_1^2 + x_2^2,$$
$$V_{12}(x_1, x_2) = x_1 x_2,$$
$$V_{21}(x_1, x_2) = x_1^2 - x_2^2,$$
$$V_{22}(x_1, x_2) = x_1 + x_2, \quad (x_1, x_2) \in \mathbb{R}^2.$$

We have

$$V_{11x_1}(x_1, x_2) = 2x_1,$$

$$V_{11x_2}(x_1, x_2) = 2x_2,$$
$$V_{12x_1}(x_1, x_2) = x_2,$$
$$V_{12x_2}(x_1, x_2) = x_1,$$
$$V_{21x_1}(x_1, x_2) = 2x_1,$$
$$V_{21x_2}(x_1, x_2) = -2x_2,$$
$$V_{22x_1}(x_1, x_2) = 1,$$
$$V_{22x_2}(x_1, x_2) = 1,$$
$$f_{x_1}(x_1, x_2) = 3x_1^2,$$
$$f_{x_2}(x_1, x_2) = 3x_2^2, \quad (x_1, x_2) \in \mathbb{R}^2.$$

Therefore,

$$\sum_{k=1}^{2} V_{2k}(x_1, x_2) \frac{\partial f}{\partial x_k}(x_1, x_2) = V_{21}(x_1, x_2) f_{t_1}(x_1, x_2) + V_{22}(x_1, x_2) f_{t_2}(x_1, x_2)$$
$$= 3x_1^2(x_1^2 - x_2^2) + 3x_2^2(x_1 + x_2)$$
$$= 3x_1^4 - 3x_1^2 x_2^2 + 3x_1 x_2^2 + 3x_2^3, \quad (x_1, x_2) \in \mathbb{R}^2.$$

Let

$$h(x_1, x_2) = 3x_1^4 - 3x_1^2 x_2^2 + 3x_1 x_2^2 + 3x_2^3, \quad (x_1, x_2) \in \mathbb{R}^2.$$

Then

$$h_{x_1}(x_1, x_2) = 12x_1^3 - 6x_1 x_2^2 + 3x_2^2,$$
$$h_{x_2}(x_1, x_2) = -6x_1^2 x_2 + 6x_1 x_2 + 9x_2^2, \quad (x_1, x_2) \in \mathbb{R}^2.$$

Hence,

$$V_1(V_2(f))(x_1, x_2) = V_{11}(x_1, x_2) h_{x_1}(x_1, x_2) + V_{12}(x_1, x_2) h_{x_2}(x_1, x_2)$$
$$= (x_1^2 + x_2^2)(12x_1^3 - 6x_1 x_2^2 + 3x_2^2) + x_1 x_2(-6x_1^2 x_2 + 6x_1 x_2 + 9x_2^2)$$
$$= 12x_1^5 - 6x_1^3 x_2^2 + 3x_1^2 x_2^2 + 12x_1^3 x_2^2 - 6x_1 x_2^4$$
$$+ 3x_2^4 - 6x_1^3 x_2^2 + 6x_1^2 x_2^2 + 9x_1 x_2^3$$
$$= 12x_1^5 + 3x_2^4 - 6x_1 x_2^4 + 9x_1^2 x_2^2 + 9x_1 x_2^3, \quad (x_1, x_2) \in \mathbb{R}^2.$$

Next,

$$V_{11}(x_1, x_2) f_{x_1}(x_1, x_2) + V_{12}(x_1, x_2) f_{x_2}(x_1, x_2) = 3x_1^2(x_1^2 + x_2^2) + 3x_2^2(x_1 x_2)$$
$$= 3x_1^4 + 3x_1^2 x_2^2 + 3x_1 x_2^3, \quad (x_1, x_2) \in \mathbb{R}^2.$$

Let

$$g(x_1, x_2) = 3x_1^4 + 3x_1^2 x_2^2 + 3x_1 x_2^3, \quad (x_1, x_2) \in \mathbb{R}^2.$$

Then

$$g_{x_1}(x_1, x_2) = 12x_1^3 + 6x_1 x_2^2 + 3x_2^3,$$
$$g_{x_2}(x_1, x_2) = 6x_1^2 x_2 + 9x_1 x_2^2, \quad (x_1, x_2) \in \mathbb{R}^2.$$

Therefore,

$$V_2(V_1(f))(x_1, x_2) = V_{21}(x_1, x_2)g_{x_1}(x_1, x_2) + V_{22}(x_1, x_2)g_{x_2}(x_1, x_2)$$
$$= (x_1^2 - x_2^2)(12x_1^3 + 6x_1 x_2^2 + 3x_2^3) + (x_1 + x_2)(6x_1^2 x_2 + 9x_1 x_2^2)$$
$$= 12x_1^5 - 6x_1^3 x_2^2 + 3x_1^2 x_2^3 - 6x_1 x_2^4 - 3x_2^5 + 6x_1^3 x_2$$
$$+ 9x_1^2 x_2^2 + 6x_1^2 x_2^2 + 9x_1 x_2^3, \quad (x_1, x_2) \in \mathbb{R}^2.$$

Exercise 6.8. Let

$$V_1(x_1, x_2) = e^{x_1^3 - x_2^2} \frac{\partial}{\partial x_1} + e^{x_1 x_2^3} \frac{\partial}{\partial x_2},$$

$$V_2(x_1, x_2) = e^{x_1^2 x_2^2} \frac{\partial}{\partial x_1} + e^{x_1 - 3x_2 + 4x_1^2} \frac{\partial}{\partial x_2},$$

$$V_3(x_1, x_2) = e^{x_1^4 + x_2^5} \frac{\partial}{\partial x_1} + e^{x_1^6 - x_2^3} \frac{\partial}{\partial x_2},$$

$$V_4(x_1, x_2) = e^{x_1^2} \frac{\partial}{\partial x_1} + e^{x_2^2} \frac{\partial}{\partial x_2},$$

$$f(x_1, x_2) = x_1 x_2^4, \quad (x_1, x_2) \in \mathbb{R}^2.$$

Prove that
1. $V_1(V_2(f)) \neq V_2(V_1(f))$;
2. $V_1(V_3(f)) \neq V_3(V_1(f))$;
3. $V_1(V_4(f)) \neq V_4(V_1(f))$;
4. $V_2(V_3(f)) \neq V_3(V_2(f))$;
5. $V_2(V_4(f)) \neq V_4(V_2(f))$;
6. $V_3(V_4(f)) \neq V_4(V_3(f))$.

Exercise 6.9. Let $V_k, k = 1, \ldots, 4$, be vector fields such that

$$V_k = \sum_{j=1}^{n} V_{kj} \frac{\partial}{\partial x_k}, \quad k = 1, \ldots, 4.$$

Let also, $a, b, c, d \in \mathbb{R}$. Prove that

$$(aV_1 + bV_2)(cV_3 + dV_4) = acV_1(V_3) + adV_1(V_4) + bcV_2(V_3) + bdV_2(V_4).$$

Solution. We have

$$aV_1 = a\left(\sum_{j=1}^{n} V_{1j}\frac{\partial}{\partial x_j}\right) = \sum_{j=1}^{n}(aV_{1j})\frac{\partial}{\partial x_j},$$

$$bV_2 = b\left(\sum_{j=1}^{n} V_{2j}\frac{\partial}{\partial x_j}\right) = \sum_{j=1}^{n}(bV_{2j})\frac{\partial}{\partial x_j},$$

$$cV_3 = c\left(\sum_{j=1}^{n} V_{3j}\frac{\partial}{\partial x_j}\right) = \sum_{j=1}^{n}(cV_{3j})\frac{\partial}{\partial x_j},$$

$$dV_4 = d\left(\sum_{j=1}^{n} V_{4j}\frac{\partial}{\partial x_j}\right) = \sum_{j=1}^{n}(dV_{4j})\frac{\partial}{\partial x_j},$$

and

$$aV_1 + bV_2 = \sum_{j=1}^{n}(aV_{1j})\frac{\partial}{\partial x_j} + \sum_{j=1}^{n}(bV_{2j})\frac{\partial}{\partial x_j}$$

$$= \sum_{j=1}^{n}(aV_{1j} + bV_{2j})\frac{\partial}{\partial x_j},$$

$$cV_1 + dV_2 = \sum_{j=1}^{n}(cV_{3j})\frac{\partial}{\partial x_j} + \sum_{k=1}^{n}(dV_{4k})\frac{\partial}{\partial x_j}$$

$$= \sum_{j=1}^{n}(cV_{3j} + dV_{4j})\frac{\partial}{\partial x_j}.$$

Hence,

$$(aV_1 + bV_2)(cV_3 + dV_4) = \sum_{j=1}^{n}(aV_{1j} + bV_{2j})\frac{\partial}{\partial x_j}\left(\sum_{j=1}^{n}(cV_{3j} + dV_{4j})\frac{\partial}{\partial x_j}\right)$$

$$= ac\sum_{j=1}^{n} V_{1j}\frac{\partial}{\partial x_j}\left(\sum_{j=1}^{n} V_{3j}\frac{\partial}{\partial x_j}\right)$$

$$+ ad\sum_{j=1}^{n} V_{1j}\frac{\partial}{\partial x_j}\left(\sum_{j=1}^{n} V_{4j}\frac{\partial}{\partial x_j}\right)$$

$$+ bc\sum_{j=1}^{n} V_{2j}\frac{\partial}{\partial x_j}\left(\sum_{j=1}^{n} V_{3j}\frac{\partial}{\partial x_j}\right)$$

$$+ bd\sum_{j=1}^{n} V_{2j}\frac{\partial}{\partial x_j}\left(\sum_{j=1}^{n} V_{4j}\frac{\partial}{\partial x_j}\right)$$

$$= acV_1(V_3) + adV_1(V_4) + bcV_2(V_3) + bdV_2(V_4).$$

This completes the proof.

Definition 6.7. Let V_1 and V_2 be vector fields. The expression

$$[V_1, V_2] = V_1(V_2) - V_2(V_1)$$

is called the Lie bracket of V_1 and V_2.

Example 6.7. Let V_1, V_2, and f be as in Example 6.6. Then

$$\begin{aligned}
[V_1, V_2] &= 12x_1^5 + 3x_2^4 - 6x_1x_2^4 + 9x_1^2x_2^2 + 9x_1x_2^3 - 12x_1^5 + 6x_1^3x_2^2 - 3x_1^2x_2^3 + 6x_1x_2^4 + 3x_2^5 \\
&\quad - 6x_1^3x_2 - 9x_1^2x_2^2 - 6x_1^2x_2^2 - 9x_1x_2^3 \\
&= 3x_2^4 + 6x_1^3x_2^2 - 3x_1^2x_2^3 + 3x_2^5 - 6x_1^3x_2 - 6x_1^2x_2^2, \quad (x_1, x_2) \in \mathbb{R}^2.
\end{aligned}$$

Exercise 6.10. Let V_1 and V_2 be vector fields. Prove that

$$[V_1, V_2] = -[V_2, V_1].$$

Exercise 6.11. Let V_1 and V_2 be vector fields. Prove that

$$[V_1, V_2] = \nabla_{V_1} V_2 - \nabla_{V_2} V_1.$$

6.5 Advanced practical problems

Problem 6.1. Find $\frac{\partial f}{\partial v}\big|_p$, where

1.
$$f(x_1, x_2) = 5x_1 + 10x_1^2x_2 + x_2^5, \quad (x_1, x_2) \in \mathbb{R}^2,$$
$$v = (4, -3), \quad p(1, 2);$$

2.
$$f(x_1, x_2, x_3) = x_1x_2^2x_3^3, \quad (x_1, x_2, x_3) \in \mathbb{R}^3,$$
$$v = (4, 3, 0), \quad p(3, 2, 1);$$

3.
$$f(x_1, x_2, x_3) = \arcsin \frac{x_3}{\sqrt{x_1^2 + x_2^2}}, \quad (x_1, x_2, x_3) \in \mathbb{R}^3, \quad (x_1, x_2) \neq (0, 0),$$
$$v = (0, 4, 3), \quad p(1, 1, 1);$$

4.
$$f(x_1, x_2, x_3, x_4) = \frac{x_2}{x_1^2 + x_2^2 + x_3^2 + x_4^2}, \quad (x_1, x_2, x_3, x_4) \in \mathbb{R}^4 \setminus (0, 0, 0, 0),$$
$$v = (3, 1, 0, 0), \quad p(0, 1, 1, 0).$$

Answer 6.4. 1. -18;
2. $\frac{52}{5}$;

3. $\frac{1}{5}$;

4. 0.

Problem 6.2. Find max $\frac{\partial f}{\partial v}\big|_p$, where

1.
$$f(x_1, x_2) = x_1 x_2^2 - 3x_1^4 x_2^5, \quad p(1,1);$$

2.
$$f(x_1, x_2) = \frac{x_1 + \sqrt{x_2}}{x_2}, \quad p(2,1);$$

3.
$$f(x_1, x_2, x_3) = \log(x_1 x_2 x_3), \quad p(1,-2,-3);$$

4.
$$f(x_1, x_2, x_3) = \tan x_1 - x_1 + 3\sin x_2 - (\sin x_2)^3 + 2x_3 + \cot x_3, \quad p\left(\frac{\pi}{4}, \frac{\pi}{3}, \frac{\pi}{2}\right).$$

Answer 6.5. 1. $\sqrt{290}$;

2. $\frac{\sqrt{29}}{2}$;

3. $\frac{7}{6}$;

4. $\sqrt{\frac{137}{8}}$.

Problem 6.3. Find an unit vector v such that $\frac{\partial f}{\partial v}$ achieves its maximum at p, where

1.
$$f(x_1, x_2) = x_1^2 - x_1 x_2 + x_2^2, \quad p(-1,2);$$

2.
$$f(x_1, x_2) = x_1 - 3x_2 + \sqrt{3x_1 x_2}, \quad p(3,1);$$

3.
$$f(x_1, x_2, x_3) = \arcsin(x_1 x_2) + \arcsin(x_2 x_3), \quad p(1,0,5,0);$$

4.
$$f(x_1, x_2, x_3) = x_1 x_3^{x_2}, \quad p(-3,2,1);$$

Answer 6.6. 1. $(-\frac{4}{\sqrt{41}}, \frac{5}{\sqrt{41}})$;

2. $(\frac{1}{\sqrt{2}}, -\frac{1}{\sqrt{2}})$;

3. $(\frac{2}{\sqrt{23}}, \frac{4}{\sqrt{23}}, -\frac{\sqrt{3}}{\sqrt{23}})$;

4. $(\frac{1}{\sqrt{37}}, 0, -\frac{6}{\sqrt{37}})$.

Problem 6.4. Using the Euler identity, find

$$x_1 \frac{\partial f}{\partial x_1} + x_2 \frac{\partial f}{\partial x_2} + x_3 \frac{\partial f}{\partial x_3},$$

where

1.
$$f(x_1, x_2) = \frac{x_1}{x_1^2 + x_2^2};$$

2.
$$f(x_1, x_2, x_3) = \frac{x_1 + x_3}{\sqrt[3]{x_1^2 + x_3^2}};$$

3.
$$f(x_1, x_2, x_3) = (x_1 + 2x_2 + 3x_3)^4;$$

4.
$$f(x_1, x_2, x_3) = (\log x_1 - \log x_2)^{\frac{x_2}{x_3}};$$

5.
$$f(x_1, x_2, x_3) = \frac{x_1 x_2}{x_3} \log x_1 + x_1 \phi\left(\frac{x_2}{x_1}, \frac{x_3}{x_1}\right),$$

where ϕ is a differentiable function.

Answer 6.7. 1.
$$-\frac{x_1}{x_1^2 + x_2^2};$$

2.
$$\frac{x_1 + x_3}{3\sqrt[3]{x_1^2 + x_3^2}};$$

3.
$$0;$$

4.
$$\frac{x_1 x_2}{x_3}(1 + \log x_1) + x_1 \phi\left(\frac{x_2}{x_1}, \frac{x_3}{x_1}\right).$$

Problem 6.5. Let

$$V_1 = (x_1^2 + x_1 x_2)\frac{\partial}{\partial x_1} + (x_1 - 2x_2)\frac{\partial}{\partial x_2},$$

$$V_2 = (x_1 + x_2^3)\frac{\partial}{\partial x_1} + 2x_2\frac{\partial}{\partial x_2}.$$

Find
1. $2V_1 + V_2$;
2. $V_1 - 3V_2$;
3. $(x_1^2 + x_2)V_1$;
4. $x_1 V_2$.

Answer 6.8. 1.
$$(2x_1^2 + 2x_1 x_2 + x_1 + x_2^3)\frac{\partial}{\partial x_1} + 2x_1\frac{\partial}{\partial x_2};$$

2.
$$(x_1^2 + x_1 x_2 - 3x_1 - 3x_2^3)\frac{\partial}{\partial x_1} + (x_1 - 8x_2)\frac{\partial}{\partial x_2};$$

3.
$$(x_1^2 + x_2)\left((x_1^2 + x_1 x_2)\frac{\partial}{\partial x_1} + (x_1 - 2x_2)\frac{\partial}{\partial t_2}\right);$$

4.
$$x_1\left((x_1 + x_2^3)\frac{\partial}{\partial x_1} + 2x_2\frac{\partial}{\partial x_2}\right).$$

Problem 6.6. Let $n = 3$ and $v = (\frac{1}{\sqrt{3}}, \frac{1}{\sqrt{3}}, \frac{1}{\sqrt{3}})$. Find $\nabla_v W$, where

1.
$$W(x_1, x_2, x_3) = (x_1 x_2 + x_2 x_3 + x_1 x_3)\frac{\partial}{\partial x_1} + \frac{1}{\sqrt{x_1^2 + x_2^2 + x_3^2}}\frac{\partial}{\partial x_2} + \left(\frac{x_3}{x_1} + \frac{x_1}{x_3}\right)\frac{\partial}{\partial x_3},$$

for $(x_1, x_2, x_3) \in \mathbb{R}^3$;

2.
$$W(x_1, x_2, x_3) = \left(\frac{x_2}{x_3} + \arcsin\left(\frac{x_3}{x_1}\right) + \arctan\left(\frac{x_1}{x_3}\right)\right)\frac{\partial}{\partial x_1} + x_3^{x_1 x_2}\frac{\partial}{\partial x_2} + \left(\frac{x_1}{x_2}\right)^{x_3}\frac{\partial}{\partial x_3},$$

for $(x_1, x_2, x_3) \in \mathbb{R}^3$.

Answer 6.9. 1.
$$\nabla_v W(x_1, x_2, x_3) = \frac{2}{\sqrt{3}}(x_1 + x_2 + x_3)\frac{\partial}{\partial x_1} - \frac{x_1 + x_2 + x_3}{\sqrt{3(t_1^2 + t_2^2 + t_3^2)}}\frac{\partial}{\partial t_2}$$
$$+ \left(\frac{1}{x_3} - \frac{x_3}{x_1^2} + \frac{1}{x_1} - \frac{x_1}{x_3^2}\right)\frac{\partial}{\partial x_3},$$

for $(x_1, x_2, x_3) \in \mathbb{R}^3$;

2.
$$\nabla_v W(x_1, x_2, x_3) = -\frac{x_1}{\sqrt{3}x_3^2}\frac{\partial}{\partial x_1} + \frac{1}{\sqrt{3}}(x_3^{x_1 x_2}(x_1 + x_2)\log x_3 + x_1 x_2 x_3^{x_1 x_2 - 1})\frac{\partial}{\partial x_2}$$
$$+ \frac{1}{\sqrt{3}}\left(\frac{x_3}{x_1} - \frac{x_3}{x_2} + \log\left(\frac{x_1}{x_2}\right)\right)\left(\frac{x_1}{x_2}\right)^{x_3}\frac{\partial}{\partial x_3},$$

for $(x_1, x_2, x_3) \in \mathbb{R}^3$.

Problem 6.7. Let
$$V_1(x_1, x_2) = e^{x_1 + 2x_2^4}\frac{\partial}{\partial x_1} + e^{x_1^3 + x_2^2}\frac{\partial}{\partial x_2},$$
$$V_2(x_1, x_2) = e^{x_1^2 + x_2^2 + x_1 x_2}\frac{\partial}{\partial x_1} + e^{x_1 x_2 + x_1^2 + x_2^3}\frac{\partial}{\partial x_2},$$
$$V_3(x_1, x_2) = e^{x_1^4 + x_2 + x_1^2}\frac{\partial}{\partial x_1} + e^{x_1^2 + x_2^4}\frac{\partial}{\partial x_2},$$

$$V_4(x_1, x_2) = e^{x_1^2 + x_1 x_2} \frac{\partial}{\partial x_1} + e^{x_1^2 + x_2^2} \frac{\partial}{\partial x_2},$$

$$f(x_1, x_2) = e^{x_1^2 + x_2^2}, \quad (x_1, x_2) \in \mathbb{R}^2.$$

Prove that

1. $V_1(V_2(f)) \neq V_2(V_1(f))$;
2. $V_1(V_3(f)) \neq V_3(V_1(f))$;
3. $V_1(V_4(f)) \neq V_4(V_1(f))$;
4. $V_2(V_3(f)) \neq V_3(V_2(f))$;
5. $V_2(V_4(f)) \neq V_4(V_2(f))$;
6. $V_3(V_4(f)) \neq V_4(V_3(f))$.

Problem 6.8. Let

$$V_1(x_1, x_2) = (x_1^4 + x_1^3 + x_2) \frac{\partial}{\partial x_1} + (e^3 x_1 - e^4 x_2^2) \frac{\partial}{\partial x_2},$$

$$V_2(x_1, x_2) = (x_1 - x_2^7) \frac{\partial}{\partial x_1} + e^5 x_1 \frac{\partial}{\partial x_2},$$

$$f(x_1, x_2) = x_1 + x_2^2, \quad (x_1, x_2) \in \mathbb{R}^2.$$

Prove that

$$[V_1, V_2](f) \neq [V_2, V_1](f).$$

Problem 6.9. Let $V_j, j = 1, 2, 3$ be vector fields and $a, b \in \mathbb{R}$. Prove that

$$[aV_1 + bV_2, V_3] = a[V_1, V_3] + b[V_2, V_3].$$

Problem 6.10. Let $V_j, j = 1, 2, 3$, be vector fields. Prove that

$$[V_1, [V_2, V_3]] + [V_2, [V_3, V_1]] + [V_3, [V_1, V_2]] = 0.$$

7 Riemannian manifolds

In this chapter introducing the notion of a manifold, we study open sets and differentiable maps on manifolds. Lie brackets and Riemannian connection acting on the set of tangent vector fields to manifolds are given. Finally, the Koszul formula is proved.

7.1 The notion of a manifold

Definition 7.1. An m-dimensional differentiable manifold is a Hausdorff space M together with a family $\{U_j\}_{j \in I}$ of subsets such that
1. $M \subseteq \bigcup_{j \in I} U_j$.
2. For any $j \in I$, there is a homeomorphism $\phi_j : U_j \to \mathbb{R}^m$ such that $\phi_j(U_j)$ is open in $\mathbb{R}^m$.
3. For $U_j \cap U_l \neq \emptyset$, $\phi_j(U_j \cap U_l)$ is open in $\mathbb{R}^m$ and

$$\phi_l \circ \phi_j^{-1} : \phi_j(U_j \cap U_l) \to \phi_l(U_j \cap U_l)$$

is differentiable for any $j, l \in I$.

Each $\phi_j, j \in I$, is called a chart and $\phi_j^{-1}, j \in I$, is referred to as a parametrization. In addition, $\phi_j(U_j), j \in I$, is said to be a parameter domain and $\{(U_j, \phi_j)\}_{j \in I}$ is said to be an atlas. The maps

$$\phi_l \circ \phi_j^{-1} : \phi_j(U_j \cap U_l) \to \phi_l(U_j \cap U_l), \quad j, l \in I$$

are said to be coordinate transformations, or transition functions.

Example 7.1. Every open subset of $\mathbb{R}^m$ is a manifold.

Example 7.2. Let

$$M = \{(x_1, \dots, x_n) \in \mathbb{R}^n : F(x_1, \dots, x_n) = 0\},$$

where $F : \mathbb{R}^n \to \mathbb{R}^m, F \in C^1(\mathbb{R}^n)$ and

$$\text{Rank}(dF) = n - m.$$

See Exercise 7.3 for the definition of dF. Then, by the implicit function theorem, there are $x_{m+1}, \dots, x_n$ such that

$$x_{m+1} = x_{m+1}(x_1, \dots, x_m),$$

$$\vdots$$

$$x_n = x_n(x_1, \dots, x_m).$$

https://doi.org/10.1515/9783111501857-007

The map

$$(x_1,\ldots,x_m) \rightarrow (x_1,\ldots,x_m,x_{m+1},\ldots,x_n)$$

is a parametrization and the map

$$(x_1,\ldots,x_n) \rightarrow (x_1,\ldots,x_m)$$

is a chart.

Exercise 7.1. Prove that the hemispheres

$$\{(x_1,x_2,x_3) \in \mathbb{S}^2 : x_j \neq 0\}, \quad j = 1,2,3,$$

define an atlas.

Hint 7.1. Use the definition for an atlas.

Definition 7.2. If $\phi_l \circ \phi_j^{-1} \in \mathcal{C}(\phi_j(U_j \cap U_l))$, for any $j,l \in I$, then M is called a topological manifold.

Definition 7.3. If $\phi_l \circ \phi_j^{-1} \in \mathcal{C}^k(\phi_j(U_j \cap U_l))$, $j,l \in I$, $k \in \mathbb{N} \cup \{\infty\}$, then M is called a $\mathcal{C}^k$-manifold.

Definition 7.4. A subset $O \subseteq M$ is said to be open if $\phi_j(O)$, $j \in I$, is open in $\mathbb{R}^m$. This defines a topology on M as the set of all open sets.

7.2 Differentiable maps

Suppose that M is an m-dimensional differentiable manifold and N is an n-dimensional differentiable manifold. Let also $f : M \rightarrow N$ be a given map.

Definition 7.5. The map f is said to be differentiable if for all charts $\phi : M \rightarrow \mathbb{R}^m$, $\psi : N \rightarrow \mathbb{R}^n$ with $f(M) \subseteq N$, the map

$$\psi \circ f \circ \phi^{-1} : \mathbb{R}^m \rightarrow \mathbb{R}^n$$

is also differentiable.

This notion of differentiability is independent on the choice of the charts ϕ and ψ.

Definition 7.6. A diffeomorphism $f : M \rightarrow N$ is defined to be a bijective map which is differentiable in both directions. Then the two manifolds M and N are said to be diffeomorphic.

Two diffeomorphic manifolds necessarily have the same dimensions because there exists no diffeomorphism between $\mathbb{R}^m$ and $\mathbb{R}^n$, $m \neq n$.

For a chart ϕ, we denote by $(u_1, \ldots, u_k)$ the coordinates of $\mathbb{R}^k$ and by $(x_1, \ldots, x_k)$ the corresponding coordinates in M. Thus, $x_j(p)$ is the function given by the jth coordinate of $\phi(p)$,

$$x_j(p) = u_j(\phi(p)).$$

Then, for a function $f : M \to \mathbb{R}$, we set

$$\left. \frac{\partial f}{\partial x_j} \right|_p = \left. \frac{\partial}{\partial u_j} (f \circ \phi^{-1}) \right|_{\phi(p)}.$$

7.3 Tangent spaces

Let M be an m-dimensional differentiable manifold and $p \in M$.

Definition 7.7 (Geometric definition). Let $\alpha : (-c, c) \subset \mathbb{R} \to M$ be a differentiable curve on M and $f : M \to \mathbb{R}$ be a function of class $C^1(M)$. The map $\alpha'(0)$ is called a tangent vector at $p \in M$, where

$$\alpha'(0)[f] = \left. \frac{d(f \circ \alpha)}{dt} \right|_{t=0}.$$

Definition 7.8 (Algebraic definition). Set

$$C^1(M) = \{f : M \to \mathbb{R} : f \text{ continuously has the first derivatives}\}.$$

A tangent vector X_p at $p \in M$ is a map $X_p : C(M) \to \mathbb{R}$ with the following two properties:
1.
$$X_p[af + bg] = aX_p[f] + bX_p[g],$$

for any $a, b \in \mathbb{R}$ and for any $f, g \in C^1(M)$;
2.
$$X_p[fg] = fX_p[g] + X_p[f]g,$$

for any $f, g \in C^1(M)$.

The set of all tangent vectors of M at p is called the tangent space of M at p, denoted by T_pM.

Example 7.3. Let $X_p \in T_pM$ and f be a constant map. We will prove that

$$X[f] = 0.$$

In fact, we have

$$X[1] = X[1 \cdot 1]$$
$$= 1 \cdot X[1] + 1 \cdot X[1]$$
$$= 2X[1].$$

Hence,

$$X[1] = 0.$$

Take $c \in \mathbb{R}$ arbitrarily. Then, using the linearity of X_p, we find

$$X[c] = X[c \cdot 1]$$
$$= c \cdot X[1]$$
$$= 0.$$

Exercise 7.2. Prove that for any tangent vector X_p, we have the following representation:

$$X_p = \sum_{j=1}^{m} X_p(x_j) \frac{\partial}{\partial x_j}\Big|_p.$$

Solution. Consider a chart $\phi : M \to U$, where, without any loss of generality, we assume that U is an open ε-ball with $\phi(p) = 0$. Hence,

$$x_1(p) = \cdots = x_m(p) = 0.$$

Let $h : U \to \mathbb{R}$ be a differentiable function and

$$f = h \circ \phi.$$

Introduce

$$h_j(y) = \int_0^1 \frac{\partial h}{\partial u_j}(ty)dt.$$

Observe that

$$\frac{\partial h}{\partial t}(ty) = \sum_{j=1}^{m} \frac{\partial h}{\partial u_j}(ty) \frac{d(tu_j)}{dt}$$
$$= \sum_{j=1}^{m} \frac{\partial h}{\partial(u_j)}(ty)u_j.$$

Hence,

$$\sum_{j=1}^{m} h_j(y)u_j = \int_0^1 \sum_{j=1}^{m} \frac{\partial h}{\partial u_j}(ty)u_j dt$$

$$= \int_0^1 \frac{\partial h}{\partial t}(ty)dt$$

$$= h(y) - h(0).$$

Since

$$f = h \circ \phi,$$

$$f_j = h_j \circ \phi,$$

$$x_j = u_j \circ \phi, \quad j \in \{1, \ldots, n\},$$

we find

$$f(q) - f(p) = \sum_{j=1}^{m} f_j(q)x_j(q).$$

Consequently,

$$\left. \frac{\partial f}{\partial x_j} \right|_q = f_j(q).$$

Now, using the properties of a tangent vector, we find

$$X(f) = X\left(f(p) + \sum_{j=1}^{m} f_j x_j \right)$$

$$= X(f(p)) + X\left(\sum_{j=1}^{m} f_j x_j \right)$$

$$= 0 + \sum_{j=1}^{m} X(f_j x_j)$$

$$= \sum_{j=1}^{m} X(f_j(p))x_j(p) + \sum_{j=1}^{m} f_j(p)X(x_j)$$

$$= \sum_{j=1}^{m} f_j(p)X(x_j)$$

$$= \sum_{j=1}^{m} \left. \frac{\partial f}{\partial x_j} \right|_p X(x_j)$$

$$= \sum_{j=1}^{m} X(x_j) \left. \frac{\partial}{\partial x_j} \right|_p (f).$$

Note that

$$\frac{\partial}{\partial x_j}(x_l) = \begin{cases} 1 & \text{if } j \neq l, \\ 0 & \text{if } j = l. \end{cases}$$

Thus,

$$\frac{\partial}{\partial x_j}, \quad j \in \{1, \ldots, m\},$$

are linearly independent. This completes the solution.

Exercise 7.3. Let $F : M \to N$ be a differentiable map, $p \in M$, and $F(p) = q \in N$. The differential of F at p is defined to be the map

$$dF|_p : T_pM \to T_qN,$$

where

$$(dF|_p(X_p))[f] = X_p[f \circ F],$$

for any $f \in C^1(N)$. For this differential, we have

$$d(G \circ F) = dG|_{F(p)} \circ dF|_p,$$

where $F : M \to N$, $G : N \to S$, or briefly

$$d(G \circ F) = dG \circ dF.$$

Solution. Let $g \in C^1(S)$. We have

$$\begin{aligned} d(G \circ F)|_p(X_p)[g] &= X_p[g \circ G \circ F] \\ &= (dF|_p(X_p))(g \circ G) \\ &= (dG|_q(dF|_p(X_p)))(g). \end{aligned}$$

This completes the solution.

Definition 7.9. A vector field X on a differentiable manifold is an association $p \to X_p \in T_pM$, $p \in M$, such that, in any chart $\phi : M \to U$ with coordinates $x_1, \ldots, x_m$, the coefficients $\xi_j : M \to \mathbb{R}$ in the representation

$$X_p = \sum_{j=1}^m \xi_j(p) \frac{\partial}{\partial x_j}\Big|_p$$

are differentiable functions.

Exercise 7.4. Let $(x^1, \dots, x^m)$ and $(y^1, \dots, y^m)$ be two coordinate systems on a manifold. Let also F be the identity. Prove that

$$\frac{\partial}{\partial x^j} = \sum_i \frac{\partial y^i}{\partial x^j} \frac{\partial}{\partial y^i}.$$

7.4 Riemannian metrics

Let M be an m-dimensional manifold and $p \in M$. Denote by $T_p^* M$ the dual space of the space $T_p M$.

Definition 7.10. The basis of $T_p^* M$ is defined by $\{dx_j\}_{j=1}^n$, where

$$dx_j|_p\left(\frac{\partial}{\partial x_i}\Big|_p\right) = \delta_{ij} = \begin{cases} 1 & \text{if } i = j, \\ 0 & \text{if } i \neq j. \end{cases}$$

Definition 7.11. Set

$$T_p M \otimes T_p M = \{a : T_p M \times T_p M \to \mathbb{R} \text{ is differentiable}\}.$$

Define a basis

$$dx_i|_p \otimes dx_j|_p, \quad i, j \in \{1, \dots, n\},$$

where

$$dx_i|_p \otimes dx_j|_p\left(\frac{\partial}{\partial x_k}\Big|_p, \frac{\partial}{\partial x_l}\Big|_p\right) = dx_i|_p\left(\frac{\partial}{\partial x_k}\Big|_p\right) dx_j|_p\left(\frac{\partial}{\partial x_l}\Big|_p\right) = \delta_{ik}\delta_{jl}.$$

For the coefficients of the representation

$$a = \sum_{i,j} a_{ij} dx_i \otimes dx_j,$$

we have the expression

$$a_{ij} = a\left(\frac{\partial}{\partial x_i}, \frac{\partial}{\partial x_j}\right).$$

Definition 7.12 (Riemannian metrics). A Riemannian metric g on M is a bilinear association $p \to g_p \in T_p M \otimes T_p M$, $p \in M$, that satisfies the following conditions:
1. $g_p(X, Y) = g_p(Y, X)$ for any $X, Y \in T_p U$;
2. $g_p(X, X) > 0$ for any $X \in T_p U, X \neq 0$;
3. The coefficients g_{ij} in the representation

$$g_p = \sum_{i,j} g_{ij} dx_i|_p \otimes dx_j|_p$$

are differential functions.

Then the pair (M, g) is called a Riemannian manifold. The Riemannian metric is also referred to as a metric tensor.

Example 7.4. The pair $(M, g) = (\mathbb{R}^n, \langle \cdot, \cdot \rangle)$ is a Riemannian manifold, where

$$(\langle \cdot, \cdot \rangle)_{ij} = \begin{pmatrix} 1 & 0 & \cdots & 0 \\ 0 & 1 & \cdots & 0 \\ \vdots & \vdots & \ddots & \vdots \\ 0 & 0 & \cdots & 1 \end{pmatrix}$$

and $\langle \cdot, \cdot \rangle$ is the standard inner product. This space is also referred as a Euclidean space denoted by $\mathbb{E}^n$.

Example 7.5. The following is an example of a Riemannian metric in matrix form

$$(g_{ij}) = \begin{pmatrix} 1 + x_1^2 & 0 & \cdots & 0 \\ 0 & 1 + x_2^2 & \cdots & 0 \\ \vdots & \vdots & \ddots & \vdots \\ 0 & 0 & \cdots & 1 + x_n^2 \end{pmatrix}.$$

A Riemannian metric g defines an inner product g_p in T_pM for any $p \in M$ and therefore the notation $\langle X, Y \rangle$ instead of $g_p(X, Y)$ is also used.

Exercise 7.5. Let $0 < b < a$. Prove that on $(0, 2\pi) \times (0, 2\pi)$,

$$(g_{ij}) = \begin{pmatrix} b^2 & 0 \\ 0 & (a + b \cos u)^2 \end{pmatrix}$$

defines a Riemannian metric.

7.5 The Riemann connection

Suppose that (M, g) is a Riemannian manifold.

Definition 7.13 (Lie bracket). Let X and Y be two differentiable vector fields on M and $f : M \to \mathbb{R}$ be a differentiable function. The Lie bracket of X and Y is defined by

$$[X, Y](f) = X(Y(f)) - Y(X(f)).$$

It is also called the Lie derivative $\mathcal{L}_X Y$ of Y in the direction X. At $p \in M$, we have

$$[X, Y]_p(f) = X_p(Y(f)) - Y_p(X(f)).$$

Suppose that X, Y, Z are differentiable vector fields on $M, f, h, \phi : U \to \mathbb{R}$ are differentiable functions, and $\alpha, \beta \in \mathbb{R}$. In what follows, we will deduce some of the properties of the Lie brackets:

1. $[\alpha X + \beta Y, Z] = \alpha[X, Z] + \beta[Y, Z].$

Proof. We have

$$
\begin{aligned}
[\alpha X + \beta Y, Z](f) &= (\alpha X + \beta Y)(Z(f)) - Z((\alpha X + \beta Y)(f)) \\
&= \alpha X(Z(f)) + \beta Y(Z(f)) - Z(\alpha X(f) + \beta Y(f)) \\
&= \alpha X(Z(f)) + \beta Y(Z(f)) - \alpha Z(X(f)) - \beta Z(Y(f)) \\
&= \alpha(X(Z(f)) - Z(X(f))) + \beta(Y(Z(f)) - Z(Y(f))) \\
&= \alpha[X, Z](f) + \beta[Y, Z](f).
\end{aligned}
$$

This completes the proof. □

2. $[X, Y] = -[Y, X].$

Proof. We have

$$
\begin{aligned}
[X, Y](f) &= X(Y(f)) - Y(X(f)) \\
&= -(Y(X(f)) - X(Y(f))) \\
&= -[Y, X](f).
\end{aligned}
$$

This completes the proof. □

3. $[fX, hY] = fh[X, Y] + fX(h)Y - hY(f)X.$

Proof. We have

$$
\begin{aligned}
[fX, hY](\phi) &= (fX)(hY(\phi)) - (hY)(fX(\phi)) \\
&= fX(hY(\phi)) - hY(fX(\phi)) \\
&= fhX(Y(\phi)) + fY(\phi)X(h) - hfY(X(\phi)) - hY(f)X(\phi) \\
&= fh(X(Y(\phi)) - Y(X(\phi))) + fX(h)Y(\phi) - hY(f)X(\phi) \\
&= fh[X, Y](\phi) + fX(h)Y(\phi) - hY(f)X(\phi).
\end{aligned}
$$

This completes the proof. □

4. $[X, [Y, Z]] + [Y, [Z, X]] + [Z, [X, Y]] = 0.$

Proof. We have

$$
\begin{aligned}
&[X, [Y, Z]](\phi) + [Y, [Z, X]](\phi) + [Z, [X, Y]](\phi) \\
&= X([Y, Z](\phi)) - ([Y, Z])(X(\phi))
\end{aligned}
$$

$$+ Y([Z,X](\phi)) - ([Z,X])(Y(\phi))$$
$$+ Z([X,Y](\phi)) - [X,Y](Z(\phi))$$
$$= X(Y(Z(\phi))) - X(Z(Y(\phi))) - Y(Z(X(\phi))) + Z(Y(X(\phi)))$$
$$+ Y(Z(X(\phi))) - Y(X(Z(\phi))) - Z(X(Y(\phi))) + X(Z(Y(\phi)))$$
$$+ Z(X(Y(\phi))) - Z(Y(X(\phi))) - X(Y(Z(\phi))) + Y(X(Z(\phi)))$$
$$= 0(\phi).$$

This completes the proof. □

5. $[\frac{\partial}{\partial x_j}, \frac{\partial}{\partial x_l}] = 0$, $j, l \in \{1, \ldots, n\}$.

Proof. We have

$$\left[\frac{\partial}{\partial x_j}, \frac{\partial}{\partial x_l} \right](\phi) = \left(\frac{\partial}{\partial x_j} \right)\left(\frac{\partial}{\partial x_l}(\phi) \right) - \left(\frac{\partial}{\partial x_l} \right)\left(\frac{\partial}{\partial x_j}(\phi) \right)$$

$$= \left(\frac{\partial}{\partial x_j} \right)\left(\frac{\partial}{\partial x_l}(\phi) \right) - \left(\frac{\partial}{\partial x_j} \right)\left(\frac{\partial}{\partial x_l}(\phi) \right)$$

$$= 0(\phi).$$

This completes the proof. □

6.
$$\left[\sum_j \xi_j \frac{\partial}{\partial x_j}, \sum_l \eta_l \frac{\partial}{\partial x_l} \right] = \sum_{j,l} \left(\xi_j \frac{\partial \eta_l}{\partial x_j} - \eta_j \frac{\partial \xi_l}{\partial x_j} \right) \frac{\partial}{\partial x_l}.$$

Proof. We have

$$\left[\sum_j \xi_j \frac{\partial}{\partial x_j}, \sum_l \eta_l \frac{\partial}{\partial x_l} \right](\phi) = \sum_j \xi_j \left(\frac{\partial}{\partial x_j}\left(\left(\sum_l \eta_l \frac{\partial}{\partial x_l} \right)(\phi) \right) \right)$$

$$- \sum_l \eta_l \left(\frac{\partial}{\partial x_l}\left(\left(\sum_j \xi_j \frac{\partial}{\partial x_j} \right)(\phi) \right) \right)$$

$$= \sum_j \xi_j \left(\sum_l \frac{\partial \eta_l}{\partial x_j} \frac{\partial \phi}{\partial x_l} + \sum_l \eta_l \frac{\partial^2 \phi}{\partial x_j \partial x_l} \right)$$

$$- \sum_l \eta_l \left(\sum_j \frac{\partial \xi_j}{\partial x_l} \frac{\partial \phi}{\partial x_j} + \sum_j \xi_j \frac{\partial^2 \phi}{\partial x_j \partial x_l} \right)$$

$$= \sum_j \xi_j \left(\sum_l \frac{\partial \eta_l}{\partial x_j} \frac{\partial \phi}{\partial x_l} \right) - \sum_l \eta_l \left(\sum_j \frac{\partial \xi_j}{\partial x_l} \frac{\partial \phi}{\partial x_j} \right)$$

$$= \sum_{j,l} \left(\xi_j \frac{\partial \eta_l}{\partial x_j} - \eta_j \frac{\partial \xi_l}{\partial x_j} \right) \frac{\partial \phi}{\partial x_l}.$$

This completes the proof. □

Definition 7.14 (Riemann connection). A Riemann connection ∇ on a Riemannian manifold $(M, \langle, \rangle)$ is a map

$$(X, Y) \to \nabla_X Y$$

that satisfies the following conditions:
1. $\nabla_{X_1 + X_2} Y = \nabla_{X_1} Y + \nabla_{X_2} Y$;
2. $\nabla_{fX} Y = f \nabla_X Y$;
3. $\nabla_X (Y_1 + Y_2) = \nabla_X Y_1 + \nabla_X Y_2$;
4. $\nabla_X (fY) = f \nabla_X Y + X(f) Y$;
5. $X \langle Y, Z \rangle = \langle \nabla_X Y, Z \rangle + \langle Y, \nabla_X Z \rangle$;
6. $\nabla_X Y - \nabla_Y X - [X, Y] = 0$.

Exercise 7.6 (Koszul formula). Prove that, for any three vector fields X, Y, Z, we have the following equation:

$$2 \langle Z, \nabla_X Y \rangle = X \langle Y, Z \rangle + Y \langle X, Z \rangle - Z \langle X, Y \rangle$$
$$- \langle Y, [X, Z] \rangle - \langle X, [Y, Z] \rangle - \langle Z, [Y, X] \rangle. \tag{7.1}$$

Solution. By Property 5, we get

$$X \langle Y, Z \rangle = \langle \nabla_X Y, Z \rangle + \langle Y, \nabla_X Z \rangle,$$
$$Y \langle X, Z \rangle = \langle \nabla_Y X, Z \rangle + \langle X, \nabla_Y Z \rangle,$$
$$-Z \langle X, Y \rangle = -\langle \nabla_Z X, Y \rangle - \langle X, \nabla_Z Y \rangle.$$

Hence,

$$X \langle Y, Z \rangle + Y \langle X, Z \rangle - Z \langle X, Y \rangle$$
$$= \langle Y, \nabla_X Z - \nabla_Z X \rangle + \langle X, \nabla_Y Z - \nabla_Z Y \rangle + \langle Z, \nabla_X Y + \nabla_Y X \rangle$$
$$= \langle Y, [X, Z] \rangle + \langle X, [Y, Z] \rangle + \langle Z, [Y, X] + 2 \nabla_X Y \rangle.$$

Using the latter equation, we obtain

$$2 \langle Z, \nabla_X Y \rangle = X \langle Y, Z \rangle + Y \langle X, Z \rangle - Z \langle X, Y \rangle$$
$$- \langle Y, [X, Z] \rangle - \langle X, [Y, Z] \rangle - \langle Z, [Y, X] \rangle.$$

This completes the proof.

Exercise 7.7. Prove that on any Riemannian manifold $(M, \langle \cdot, \cdot \rangle)$ there is a uniquely determined Riemann connection.

Solution. 1. Uniqueness. Note that for a given Z, the right-hand side of (7.1) is uniquely determined. Hence, $\nabla_X Y$ is uniquely determined.

2. Existence. We define ∇ by the equality (7.1). We will check that it satisfies all the requirements of Definition 7.14.

a. We have

$$2\langle Z, \nabla_{X_1} Y \rangle = X_1 \langle Y, Z \rangle + Y \langle X_1, Z \rangle - Z \langle X_1, Y \rangle$$
$$- \langle Y, [X_1, Z] \rangle - \langle X_1, [Y, Z] \rangle - \langle Z, [Y, X_1] \rangle,$$

$$2\langle Z, \nabla_{X_2} Y \rangle = X_2 \langle Y, Z \rangle + Y \langle X_2, Z \rangle - Z \langle X_2, Y \rangle$$
$$- \langle Y, [X_2, Z] \rangle - \langle X_2, [Y, Z] \rangle - \langle Z, [Y, X_2] \rangle,$$

$$2\langle Z, \nabla_{X_1} Y \rangle + 2\langle Z, \nabla_{X_2} Y \rangle = X_1 \langle Y, Z \rangle + Y \langle X_1, Z \rangle - Z \langle X_1, Y \rangle$$
$$- \langle Y, [X_1, Z] \rangle - \langle X_1, [Y, Z] \rangle - \langle Z, [Y, X_1] \rangle$$
$$+ X_2 \langle Y, Z \rangle + Y \langle X_2, Z \rangle - Z \langle X_2, Y \rangle$$
$$- \langle Y, [X_2, Z] \rangle - \langle X_2, [Y, Z] \rangle - \langle Z, [Y, X_2] \rangle$$

and

$$2\langle Z, \nabla_{X_1+X_2} Y \rangle = (X_1 + X_2)\langle Y, Z \rangle + Y \langle X_1 + X_2, Z \rangle - Z \langle X_1 + X_2, Y \rangle$$
$$- \langle Y, [X_1 + X_2, Z] \rangle - \langle X_1 + X_2, [Y, Z] \rangle - \langle Z, [Y, X_1 + X_2] \rangle$$
$$= X_1 \langle Y, Z \rangle + X_2 \langle Y, Z \rangle + Y \langle X_1, Z \rangle + Y \langle X_2, Z \rangle$$
$$- Z \langle X_1, Y \rangle - Z \langle X_2, Y \rangle - \langle Y, [X_1, Z] \rangle - \langle Y, [X_2, Z] \rangle$$
$$- \langle X_1, [Y, Z] \rangle - \langle X_2, [Y, Z] \rangle - \langle Z, [Y, X_1] \rangle - \langle Z, [Y, X_2] \rangle.$$

Consequently,

$$2\langle Z, \nabla_{X_1+X_2} Y \rangle = 2\langle Z, \nabla_{X_1} Y \rangle + 2\langle Z, \nabla_{X_2} Y \rangle,$$

and from here

$$\nabla_{X_1+X_2} Y = \nabla_{X_1} Y + \nabla_{X_2} Y.$$

b. We have

$$2\langle Z, \nabla_{fX} Y \rangle = fX \langle Y, Z \rangle + Y \langle fX, Z \rangle - Z \langle fX, Y \rangle$$
$$- \langle Y, [fX, Z] \rangle - \langle fX, [Y, Z] \rangle - \langle Z, [Y, fX] \rangle$$
$$= fX \langle Y, Z \rangle + Y(f) \langle X, Z \rangle + fY \langle X, Z \rangle$$
$$- Z(f) \langle X, Y \rangle - fZ \langle X, Y \rangle$$
$$- \langle Y, f[X, Z] - Z(f)X \rangle - f\langle X, [Y, Z] \rangle$$
$$- \langle Z, -Y(f)X + f[X, Y] \rangle$$
$$= fX \langle Y, Z \rangle + Y(f) \langle Z, X \rangle + fY \langle X, Z \rangle$$
$$- Z(f) \langle X, Y \rangle - fZ \langle X, Y \rangle$$
$$- f\langle Y, [X, Z] \rangle + Z(f) \langle Y, X \rangle - f\langle X, [Y, Z] \rangle$$

$$-Y(f)\langle Z, X\rangle - f\langle Z, [X, Y]\rangle$$
$$= f(X\langle Y, Z\rangle + Y\langle X, Z\rangle - Z\langle X, Y\rangle$$
$$- \langle Y, [X, Z]\rangle - \langle X, [Y, Z]\rangle - \langle Z, [X, Y]\rangle)$$

and

$$\nabla_{fX} Y = f\nabla_X Y.$$

c. We have

$$2\langle Z, \nabla_X(Y_1 + Y_2)\rangle = X\langle Y_1 + Y_2, Z\rangle + (Y_1 + Y_2)\langle X, Z\rangle - Z\langle X, Y_1 + Y_2\rangle$$
$$- \langle Y_1 + Y_2, [X, Z]\rangle - \langle X, [Y_1 + Y_2, Z]\rangle - \langle Z, [Y_1 + Y_2, X]\rangle$$
$$= X\langle Y_1, Z\rangle + Y_1\langle X, Z\rangle - Z\langle X, Y_1\rangle$$
$$+ X\langle Y_2, Z\rangle + Y_2\langle X, Z\rangle - Z\langle X, Y_2\rangle$$
$$- \langle Y_1, [X, Z]\rangle - \langle X, [Y_1, Z]\rangle - \langle Z, [Y_1, X]\rangle$$
$$- \langle Y_2, [X, Z]\rangle - \langle X, [Y_2, Z]\rangle - \langle Z, [Y_2, X]\rangle$$

and

$$\nabla_X(Y_1 + Y_2) = \nabla_X Y_1 + \nabla_X Y_2.$$

d. We have

$$2\langle Z, \nabla_X fY\rangle = X\langle fY, Z\rangle + fY\langle X, Z\rangle - Z\langle X, fY\rangle$$
$$- \langle fY, [X, Z]\rangle - \langle X, [fY, Z]\rangle - \langle Z, [fY, X]\rangle$$
$$= X(f)\langle Y, Z\rangle + fX\langle Y, Z\rangle + fY\langle X, Z\rangle$$
$$- Z(f)\langle X, Y\rangle - fZ\langle X, Y\rangle$$
$$- f\langle Y, [X, Z]\rangle + Z(f)\langle X, Y\rangle - f\langle X, [Y, Z]\rangle$$
$$- f\langle Z, [Y, X]\rangle + X(f)\langle Y, Z\rangle$$
$$= 2X(f)\langle Y, Z\rangle + f(X\langle Y, Z\rangle + Y\langle X, Z\rangle$$
$$- Z\langle X, Y\rangle - \langle Y, [X, Z]\rangle - \langle X, [Y, Z]\rangle - \langle Z, [Y, X]\rangle)$$

and

$$\nabla_X fY = X(f)Y + f\nabla_X Y.$$

e. We have

$$2\langle Z, \nabla_X Y\rangle + 2\langle Y, \nabla_X Z\rangle = X\langle Y, Z\rangle + Y\langle X, Z\rangle - Z\langle X, Y\rangle - \langle Y, [X, Z]\rangle$$
$$- \langle X, [Y, Z]\rangle - \langle Z, [Y, X]\rangle + X\langle Z, Y\rangle + Z\langle X, Y\rangle$$
$$- Y\langle X, Z\rangle - \langle Z, [X, Y]\rangle - \langle X, [Z, Y]\rangle - \langle Y, [Z, X]\rangle$$
$$= 2X\langle Y, Z\rangle$$

and

$$X\langle Y, Z\rangle = \langle Z, \nabla_X Y\rangle + \langle Y, \nabla_X Z\rangle.$$

f. We have

$$
\begin{aligned}
2\langle Z, \nabla_X Y\rangle - 2\langle Z, \nabla_Y X\rangle &= X\langle Y,Z\rangle + Y\langle X,Z\rangle - Z\langle X,Y\rangle - \langle Y,[X,Z]\rangle \\
&\quad - \langle X,[Y,Z]\rangle - \langle Z,[Y,X]\rangle - Y\langle X,Z\rangle - X\langle Y,Z\rangle \\
&\quad + Z\langle Y,X\rangle + \langle X,[Y,Z]\rangle + \langle Y,[X,Z]\rangle + \langle Z,[X,Y]\rangle \\
&= 2\langle Z,[X,Y]\rangle
\end{aligned}
$$

and

$$\nabla_X Y - \nabla_Y Z = [X, Y].$$

This completes the proof.

7.6 The Christoffel coefficients

Set $\partial_j = \frac{\partial}{\partial x_j}$. We will find a representation of

$$\nabla_{\partial_j}\partial_l$$

in the form

$$\nabla_{\partial_j}\partial_l = \sum_k \Gamma_{jl}^k \partial_k.$$

By the Koszul formula, we get

$$
\begin{aligned}
2\langle \nabla_{\partial_j}\partial_l, \partial_k\rangle &= \partial_j\langle \partial_l, \partial_k\rangle + \partial_l\langle \partial_j, \partial_k\rangle - \partial_k\langle \partial_j, \partial_l\rangle \\
&= \frac{\partial g_{lk}}{\partial x_j} + \frac{\partial g_{jk}}{\partial x_l} - \frac{\partial g_{jl}}{\partial x_k}.
\end{aligned}
$$

Therefore,

$$\langle \nabla_{\partial_j}\partial_l, \partial_k\rangle = \frac{1}{2}\left(\frac{\partial g_{lk}}{\partial x_j} + \frac{\partial g_{jk}}{\partial x_l} - \frac{\partial g_{jl}}{\partial x_k} \right).$$

Let

$$\Gamma_{jl,k} = \frac{1}{2}\left(\frac{\partial g_{lk}}{\partial x_j} + \frac{\partial g_{jk}}{\partial x_l} - \frac{\partial g_{jl}}{\partial x_k} \right).$$

Thus, we have

$$\Gamma_{jl,k} = \langle \nabla_{\partial_j} \partial_l, \partial_k \rangle$$

$$= \left\langle \sum_m \Gamma_{jl}^m \partial_m, \partial_k \right\rangle$$

$$= \sum_m \Gamma_{jl}^m \langle \partial_m, \partial_k \rangle$$

$$= \sum_m \Gamma_{jl}^m g_{mk}.$$

Let

$$g^{km} = (g_{km})^{-1}.$$

Then

$$\Gamma_{jl}^m = \sum_k \Gamma_{jl,k} g^{km}.$$

Definition 7.15. The symbols $\Gamma_{ij,l}$ and Γ_{ij}^k are called the Christoffel coefficients.

Let

$$X = \sum_i \xi_i \partial_i,$$

$$Y = \sum_j \eta_j \partial_j.$$

Then

$$\nabla_X Y = \sum_i \xi_i \nabla_{\partial_i} \left(\sum_j \eta_j \partial_j \right)$$

$$= \sum_i \sum_j \xi_i \nabla_{\partial_i} (\eta_j \partial_j)$$

$$= \sum_i \sum_j \xi_i \eta_j \nabla_{\partial_i} \partial_j + \sum_i \sum_j \xi_i \left(\frac{\partial \eta_j}{\partial x_i} \partial_j \right)$$

$$= \sum_i \sum_j \sum_l \xi_i \eta_j \Gamma_{ij}^l \partial_l + \sum_i \sum_j \xi_i \frac{\partial \eta_j}{\partial x_i} \partial_j$$

$$= \sum_i \sum_l \sum_j \xi_i \eta_l \Gamma_{il}^j \partial_j + \sum_i \sum_j \xi_i \frac{\partial \eta_j}{\partial x_i} \partial_j$$

$$= \sum_j \left(\sum_i \xi_i \frac{\partial \eta_j}{\partial x_i} + \sum_{i,l} \xi_i \eta_l \Gamma_{il}^j \right) \partial_j.$$

Definition 7.16. A vector field Y is said to be parallel if

$$\nabla_X Y = 0$$

for any X.

Definition 7.17. A vector field Y along a regular curve c is said to be parallel along c if

$$\nabla_{c'} Y = 0.$$

Definition 7.18. A regular curve c is said to be geodesic if

$$\nabla_{c'} c' = 0.$$

7.7 Advanced practical problems

Problem 7.1. Prove that the Cartesian product $M_1 \times M_2$ of two differentiable manifolds is a differentiable manifold.

Hint 7.2. Use the definition of a differentiable manifold.

Problem 7.2. Let M be a given differentiable manifold. Prove that the set of all pairs $(p, X), X \in T_p M$, is a differentiable manifold. It is called the tangent bundle of M, denoted by TM.

Hint 7.3. For any chart ϕ in M, consider

$$\Phi(p, M) = (\phi(p), \xi^1(p), \dots, \xi^n(p)) \in \mathbb{R}^n \times \mathbb{R}^n,$$

where $\xi^1, \dots, \xi^n$ are the components of X in the corresponding basis, i. e.,

$$X_p = \sum_{j=1}^{n} \xi^j(p) \frac{\partial}{\partial x^j}.$$

Problem 7.3. Let

$$M = \{(x_1, x_2, x_3, x_3) \in \mathbb{R}^4 : x_1^2 + x_2^2 = x_3^2 + x_4^2 = 1\}.$$

Prove that M is a two-dimensional manifold.

Problem 7.4. Prove that the metrics of two Riemannian manifolds (M_1, g_1) and (M_2, g_2) induce a Riemannian metric $g_1 \times g_2$ on $M_1 \times M_2$.

Problem 7.5. In $\mathbb{R}^3$, set

$$\nabla_X Y = D_X Y + \frac{1}{2}(X \times Y).$$

Check if

$$\nabla_X Y - \nabla_Y X = [X, Y].$$

Answer 7.4. No.

Problem 7.6. Prove that the Poincaré upper half-plane

$$\{(x, y) \in \mathbb{R}^2 : y > 0\}$$

with the metric

$$(g_{ij}(x, y)) = \frac{1}{y^2} \begin{pmatrix} 1 & 0 \\ 0 & 1 \end{pmatrix}$$

is a Riemannian manifold.

Index

https://doi.org/10.1515/9783111501857-008

www.ingramcontent.com/pod-product-compliance
Lightning Source LLC
Chambersburg PA
CBHW061345210326
41598CB00035B/5889